The Achievement of
GALILEO

EDITED WITH NOTES BY
JAMES BROPHY
AND
HENRY PAOLUCCI

WITH AN INTRODUCTION BY
HENRY PAOLUCCI

PREFACE TO NEW EDITION BY
Anne Paolucci

Copyright © 2001 by Anne Paolucci

First published by Twayne Publishers, Inc.,
reprinted by College and University Press, 1962

Library of Congress Cataloging-in-Publication Data

Galilei, Galileo, 1564-1642
   The achievement of Galileo / edited with notes by James Brophy
and Henry Paolucci ; with an introduction by Henry Paolucci.
--2 nd ed. / preface to new edition by Anne Paolucci.
      p. cm.
   A selection of writings and documents by and about Galileo.
   ISBN 0-918680-95-6 (alk. paper)
   1. Galilei, Galileo, 1564-1642. I. Brophy, James D.
   II. Paolucci, Henry. Title.

QB36.G2 A233 2001
520'.92--dc21

                                        2001033444

Published by
THE BAGEHOT COUNCIL
Distributed by
GRIFFON HOUSE PUBLICATIONS
P. O. BOX 468
SMYRNA, DE 19977

Contents

Preface to New Edition

First published in 1962, *The Achievement of Galileo* reflects Professor Paolucci's life-long interest in the history of science and astronomy. He was already familiar with Greek writings and later theories on the movement of celestial bodies when he volunteered for the U.S. Air Force, in 1942. As a navigator, flying missions from North Africa into Germany and Italy (often at night), and later stationed in Italy, he had opportunity to observe and study the stars and other celestial bodies. Back in the States, his continuing interest in astronomy led him to Pierre Duhem, the French historian of science. He read (and came to own) Duhem's massive ten volumes of *Le Système du Monde, Histoire des Doctrines Cosmologiques de Platon à Copernic* and explored again other writings on the subject, especially the more recent ones, down to Ernst Mach, Einstein, and beyond.

The Galilean controversy sparked a special interest in him. Relentless in pointing out that the notorious *eppur si muove* was apocryphal, he was even more determined to challenge the notion that Galileo was right in defending the Copernican theory as "true." Following Duhem's lead, he traced the argument down to our own time and came to the conclusion that Cardinal Bellarmine's position, not Galileo's, accorded with the now universally accepted Einsteinian point of view. The excerpts chosen for this anthology — some of them translated for the first time — offer an impressive critical "dialogue" of the controversy.

In his Introduction, Professor Paolucci examines briefly Galileo's views against the uncertainties of modern science, citing writers like Philipp Frank, Morris R. Cohen, Einstein himself and others, and offers this premise:

> while twentieth-century science fully acknowledges its debt to Galileo for its mathematical-experimental method, it has unequivocally decided against him in his defense of the Copernican system.

Professor Paolucci's Conclusion sums up the arguments of this brief but important survey that helps redefine the parameters of Galileo's lasting contribution to science and astronomy. His closing words are worth quoting for their carefully articulated historical assessment; but they also serve as a striking reminder of the integrity of the scholar who was able to do justice to both the greatness and the limits of Galileo's achievement.

> In spite of Kepler and Galileo, we today hold with Osiander and Bellarmine that the hypotheses of physics are but mathematical artifices designed to *save the phenomena*, but thanks to Kepler and Galileo. we now call upon them to *save, as a single whole, all of the phenomena of the inanimate universe.*

ANNE PAOLUCCI
May 15, 2001

Introduction: Twentieth-Century Questions

It is a curious fact that, just when the man in the street has begun to believe thoroughly in science, the man in the laboratory has begun to lose his faith.

— BERTRAND RUSSELL
(from *The Scientific Outlook*)

Natural science has come a long way since that dawn of self-confidence in the opening decades of the seventeenth century when Galileo, by joining mathematics with experimentation, and by defending their union with convincing eloquence, opened up to a "vast and most excellent science," as he himself called it, "ways and means by which other minds more acute than mine will explore its remote corners." In the intervening centuries, remote corners of that science have, in fact, been explored by acute minds, many of them far better trained in mathematics and far better equipped for systematic experimentation than Galileo was; so that, quantitatively considered, the scientific knowledge he possessed amounts to but a fragment of what is available to any good physicist today. And yet, as Albert Einstein has remarked, the ways and means of advance have indeed been those conceived and perfected by Galileo. He it was who first saw and then was able to convince men of science that all knowledge of the real world of space and time must start with and end in experience, that ideas arrived at by pure thinking, without observation, are completely empty as regards reality; he it was who first insisted that the source of experience, the "great book of Nature," is a book written in mathematical characters, and that, therefore, only minds adequately trained in mathematics can so much as begin to read it intelligently. "Because Galileo saw this," Einstein has written, "and particularly because he drummed it into the scientific world, he is the father of modern physics—indeed, of modern science altogether."

But, while continuing to adhere strictly to his ways and means, modern mathematical physicists, for the most part, are far less confident about results than Galileo was. His earliest writings reveal that, even before he became a Copernican, Galileo had been convinced, as Descartes was later to be, that ultimate material reality is essentially mathematical, and that what is true of mathematical objects, therefore, is equally true of material objects. Thus, for instance, he could say: "if you had a perfect sphere and plane, though they were material, you need not doubt that they would touch at one sole point. . . . Nay, pursuing the question with more subtle contemplations, you would find that it is much harder to procure two bodies that touch with part of their surfaces than with one point only." The argument of these words is typical of Galileo's lifelong polemic against the Aristotelian scientists of his day who believed that, by applying mathematical techniques and reasoning to the study of natural phenomena, one could acquire, at best, only abstract, relative knowledge, convenient, like mnemonic devices, for summarizing facts and "saving appearances," but inadequate for comprehending the essential reality of material things. Galileo believed, on the contrary, that precisely such mathematical study alone enables the human intellect to break through appearances and gain true knowledge of the underlying reality of the empirical world. God, he did not hesitate to say, may know infinitely more than we can ever know; but what we know mathematically we know as well as He does.

This confidence of Galileo in the certitude of knowledge acquired by mathematical and experimental study served, in his own day and for centuries thereafter, to attract many studious minds away from other books, over which there was much disputing, to the book of nature. But today, in the era of post-Planckian and post-Einsteinian physics, few competently trained persons can pretend to enjoy such confidence. "Mathematics and experiment," Professor Tobias Dantzig has recently written, "reign more firmly than ever over the new physics, but an all-pervading scepticism has affected their validity." And Bertrand Russell, to the same effect, has asserted: "Science, which began as the pursuit of truth, is becoming incompatible with veracity, since complete veracity tends more and more to complete scientific scepticism."

Many thoughtful persons of our time have welcomed the change in attitude that has made modern physicists "humble and stammering" where their predecessors, heirs to the confidence of Galileo and Newton, were "proud and dictatorial." But the best scientists, the Plancks and Einsteins, have not welcomed the change. While admitting that an "unbridgeable chasm" seems, for the present, to separate natural reality, with its infinite and perhaps ultimately indeterminate complexity, and the abstract representations of scientific formulas, they refuse to resign themselves to the present state of things. "Some physicists," Einstein wrote, "among them myself, cannot believe that we must abandon, actually and forever, the idea of direct representation of physical reality in space and time; or that we must accept the view that events in nature are analogous to a game of chance." But there is, clearly, a ring of nostalgia rather than of prophecy in this Einsteinian protest. And it could hardly have been otherwise with the man whose own cultural importance as a scientist consists largely in his having shaken the confidence of dogmatists by reducing to the humble status of convenient hypotheses the great Galilean and Newtonian principles of terrestrial and celestial mechanics which, for more than two hundred years, had been held to be true in an absolute sense.

Rare is the educated man today who would insist, against Einstein, that Newton's principle of inertia, which posits an unobservable state of affairs, is anything more than an assumption, valid, scientifically, only to the extent that it introduces simplicity and economy in mathematical representations of the complicated phenomena of motion; but there are still many who, in spite of Einstein, persist in believing that the Copernican system defended by Galileo has been established as "true" in an absolute sense. Giorgio de Santillana, for instance, in his popular book *The Crime of Galileo,* says that, if Galileo himself was unable to provide "conclusive physical proof" of the new system, it was simply because he "could not yet produce Foucault's pendulum." Santillana is, of course, thoroughly aware of the revolution effected by Einstein and has frequently expressed his full accord with its results, so that it is difficult to understand why he is unable to grasp its implications with regard to the Foucault and other experiments designed to "prove" that the earth is indeed in motion. Morris R. Cohen, on the contrary, in an essay illustrating the significance of the Einsteinian revolution, has stated its implications with regard to all such proofs in these emphatic terms:

> The reader who knows something of the history of science will recognize that our example shows Einstein's later theory of relativity as reopening the issue between Galileo and those who condemned him for saying that the earth *is* in motion. . . . It would be vain to repeat against Einstein the old arguments for the absolute rotation of the earth, based on Foucault's pendulum or the bulging of the earth at the equator. He shows that it is possible to define a space with regard to which the fixed stars are rotating. In such a space the earth may be considered at rest, and the phenomena which in Newtonian mechanics are called gravitational and centrifugal would change places. Since both are proportional to the mass of the earth there would be no experimental difference.

Professor Cohen has here merely paraphrased words of Einstein himself which have become commonplaces in popular expositions of his thought, but which are nevertheless precise.

Philipp Frank, equally emphatic in his exposition, is perhaps more explicit when he writes:

> From Einstein's principles one could derive the description of the motions of celestial bodies relative to any system of reference. One could demonstrate that the description of the motion of planets becomes particularly simple if one uses the system of fixed stars as a system of reference, but there was still no objection to using the earth as system of reference. In this case, one obtains a description in which the earth is at rest and the fixed stars are in a rotational motion. What appears to be in the Copernican heliocentric system the centrifugal force of the rotating earth becomes in the geocentric system a gravitational effect of the rotating fixed stars upon the earth.
>
> The Copernican system became for the first time in its history not only mathematically but also philosophically true. But at the same moment the geocentric system became philosophically true, also. The system of reference had lost all philosophic meaning. For each astronomical problem, one had to pick the system of reference that rendered the simplest description of the motions of the celestial bodies involved.

Thus one must conclude that while twentieth-century science fully acknowledges its debt to Galileo for its mathematical-experimental method, it has unequivocally decided against him in his defense of the Copernican system. The dogmatism of the apocryphal *eppur si muove,* with which his case was originally appealed to posterity, has been repudiated by the best modern scientists; and with the popularization of *relativity,* their repudiation of it is rapidly filtering down into non-scientific circles.

Yet, when we turn from writings about Galileo to read for ourselves the text of his defense of the Copernican system, we are bound—if we have not been forewarned—to be startled by what we find in its opening pages. For nothing could be more "modern," nothing could be more consistent with the post-Einsteinian attitude of modesty toward scientific knowledge, than the preface which Galileo himself provided for his *Dialogue on the Great World Systems*. Galileo thus summarizes what he means to accomplish in his dialogues:

> Three main topics are to be treated. First, I will endeavor to show that all experiments that can be made upon the Earth are insufficient means to conclude for its mobility but are indifferently applicable to the Earth, movable or immovable; and I hope that on this occasion many observations will come to light that were unknown to the ancients. Secondly, we will examine the celestial phenomena that make for the Copernican hypothesis, as if it were to prove absolutely victorious, adding by the way certain new observations which yet serve only for astronomical facility, not for natural necessity. In the third place, I will propose an ingenious fancy . . . that the unknown problem of the tides might receive some light, admitting the Earth's motion.

Galileo is here saying in popular language precisely what Cohen and Frank have said in their expositions of the implications of Einstein's later theory of relativity: that it cannot be demonstrated experimentally whether the earth is in motion or not; that preference of one system over another is a matter of "astronomical facility," not "natural necessity"; that in the case of a particular problem, like explaining the tides, one assumption may prove to be more "convenient" than another. And, if we turn from the opening to the closing pages of the *Dialogue,* we find an even more explicit avowal of a scientific relativism: the famous interlocutors, Simplicio, Sagredo, and Salviati agree there that identical observable phenomena may be brought about, or accounted for, in diverse ways, many of them beyond the capacity of the human intellect to comprehend.

The trouble with the "relativism" which we find in the opening and concluding pages is that, as the scholars assure us, Galileo did not mean it. He was persuaded that, with his many experiments, and particularly with his excitingly new telescopic observations of the moon, sun, planets and stars, he had added to the mathematical exactitude of the Copernican system sufficient empirical support to establish it as absolutely true. The relativistic modesty of his preface and conclusion, all scholars agree, was forced upon him by ecclesiastical critics who would not otherwise have permitted the work to be published.

A number of interesting questions force themselves upon our attention at this point. How did it happen that the opponents of Galileo were able to anticipate and favor, in the seventeenth century, an attitude which the disciples of Galilean science were not to assume until the time of Ernst Mach, Hertz, Poincaré and Einstein? When they forced him to recant what he had always believed, when they forced him to profess solemnly on his knees the view he had pretended to uphold in the preface and conclusion of his *Dialogue*, were the ecclesiastical opponents of Galileo, paradoxically, forcing him to be scientifically right in spite of himself? Such questions surely are implied in Professor Cohen's assertion that Einstein, with his later theory of relativity, has reopened the issue "between Galileo and those who condemned him for saying that the earth *is* in motion."

But as we read more deeply into the great dialogues of Galileo, other, even more important questions are apt to occur to us. Was Galileo really a precursor of Newton in formulating the law of inertia? He has been called by some the "creator" of that law, though, as Santillana remarks, "he could not help recoiling from the full formulation of his own inertial principle." Yet when we turn to the text, we find him, in fact, upholding the very antithesis of the Newtonian formulation. Like the modern adherents of the field-theory, Galileo held it to be impossible "that any movable body can have a natural principle of moving in a straight line." Contrary to Newton's view that all curved motions are compounded of rectilinear motions, and quite in accord with the twentieth-century view, he believed that all apparently rectilinear motions are either actually curved or compounded of curved motions. Indeed it was primarily because the Copernican system "ennobled" the earth by attributing to it the "effortless motions" of a heavenly body that Galileo was drawn—so he tells us—to its defense.

But the reader must experience for himself the surprises that are in store when one turns from secondary sources to the actual text of Galileo's work. It should be noted that the dialogue-form which he used in his major works, is, when well used, a marvelous vehicle for surprise: if it is not to lapse into a monologue, divergent positions on the matters discussed must be given eloquent and consistent expression. Thus, for dramatic effect the author may unwittingly give utterance to arguments which he himself has rejected as false, but which may appear startlingly fresh and true to readers of subsequent generations. For readers like ourselves, in the post-Einsteinian era, this is especially true.

The selection of writings and documents by and about Galileo which is offered here is intended primarily to stimulate and in part, satisfy new interest in his thought at the same time that it facilitates reappraisal of his significance as a founder of modern science. The first chapter presents excerpts from a variety of his writings, illustrating his extraordinary capacity to convey through written words the excitement of a "watcher of the skies when a new planet swims into his ken," as well as his flair for literary controversy and the stylistic artistry of his dialogues—artistry that has earned him, in the judgment of the best Italian critics, a leading place among the master prose writers of the Italian language.

The second chapter, reviewing Galileo's involvement in the controversy over the "new" astronomy, includes the important letter of Cardinal Bellarmine advising Galileo to maintain a "relativist" attitude in his defense of Copernicus, some notes of Galileo in response to Bellarmine's advice, an early defense of Galileo by the remarkable monk, Thomas Campanella, long excerpts from Galileo's great dialogue on the subject, and, of course, the celebrated recantation which the Church authorities exacted from the penitent Galileo.

The selections in the third chapter, drawn exclusively from what he himself considered to be his masterwork, the *Dialogues Concerning Two New Sciences,* illustrate the uncontested achievement of Galileo in perfecting the method of mathematico-experimental science and in extending its application to the study of the physical phenomena of the immediate environment of man.

The fourth and final chapter surveys the reputation of Galileo through subsequent centuries. The first reading, from the famous *Encyclopédie* of the French *philosophes,* sums up, by means of lengthy citations, the unsettled state of Galileo's reputation to that time. In the closing decades of the seventeenth century and through the middle of the eighteenth, while the anti-experimental *raison raisonant* of the French *philosophes* reigned over the intellectual salons of Europe, Galileo's fame was limited to a rather narrow circle of enthusiasts. Recognition by the Encyclopedists did not come until Newton's triumph had discredited the extreme rationalism of the Cartesians; but when it came, finally, it assured universal distinction for Galileo. Thomas Henri Martin's *Galilée: Les Droits de la Science et la Méthode des Sciences Physiques,* from which the second reading is drawn, is generally acknowledged to be the first full-scale attempt by a qualified scholar to appraise, historically and scientifically, the significance of Galileo's labors both as a methodologist and as a champion of the rights of science. The reading from Andrew D. White's well-known work represents the Galileo "case" according to the established view, as a major instance of the baneful effects of the age-old conflict between religious bigotry and science. These three readings, which provided the traditional picture in its manifold aspects, are followed by selections from twentieth-century authors— Albert Einstein, Morris R. Cohen, Philipp Frank, and Pierre Duhem—which show, as we have outlined in the preceding pages, how modern physics, advancing beyond the principles of motion and the ideas of space and time of the traditional

Newtonian mechanics, has raised many new questions and has stirred up controversial interest in some very old questions regarding the scientific achievement of Galileo as an originator of the scientific method and as a champion of the Copernican astronomy. The questions these authors raise are very clear; and if their answers are not wholly satisfactory as answers, they will surely serve, at least, to stimulate new interest in Galileo's thought, and perhaps further inquiry into the controversy that seems destined to accompany his name through all the ages.

The Achievement of Galileo

CHAPTER *1*

Discovery and the Burden of the Past

NOT A SMALL PART of the great influence Galileo exerted on his own and subsequent generations was due to his masterful literary style as a polemicist. The best Italian literary critics rank him with Machiavelli and Leopardi as one of the finest Italian prose writers. De Sanctis has written:

> In Galileo there is much to remind one of Machiavelli. . . . Reading his letters, treatises and dialogues we at once become conscious of the stamp of Tuscan culture in its maturity; a style as full of things themselves as it is of thought, shorn of pretensions and mannerisms, in a directness and frankness of form which is the ultimate perfection of prose. He uses the servile terms of the courtly life of his day, but without servility; indeed his obeisances are permeated with a dignity and simplicity that lifts him above his protectors. He seeks neither elegance nor delicate refinement, writing always with clarity and rigor, like a man intent upon the substance of things and indifferent to external trappings.

In this first chapter are included samples of his literary brilliance in its three characteristic manifestations. The first selection, excerpted from his *Starry Messenger* (1610), the work that at once made him famous throughout Europe, illustrates his remarkable, Olympian capacity to convey the excitement of discovery; the second, taken from his *Assayer* (1623), illustrates his mastery of the polemical style, pursuing adversaries on their own grounds ascending as high, or descending as low as they care to carry the fight; the third, excerpted from his celebrated *Dialogue on the Great World Systems* (1632), illustrates his capacity to rise above mere polemics to genuine philosophic dialectic, in a masterful literary creation modeled on the Socratic dialogues of Plato. Whether announcing discoveries, combating critics, or recreating imaginatively, for others, the new world of science lately discovered, Galileo is a master rhetorician in the best sense of the world. Indeed, as examples of serious yet charmingly urbane intellectual discourse, the great dialogues of Galileo vie in excellence with the Plato prototypes themselves.

In addition to illustrating his literary style the three selections included in this chapter serve to introduce the reader to the chief intellectual concerns of Galileo, as a keen observer of nature, liberator of minds from the dead weight of authority, and champion of the mathematico-experimental method of scientific inquiry.

from

The Starry Messenger*

Revealing great, unusual, and
remarkable spectacles, opening these
to the consideration of every man,
and especially of philosophers
and astronomers;
AS OBSERVED BY GALILEO GALILEI
Gentleman of Florence
Professor of Mathematics in the
University of Padua,
WITH THE AID OF A
SPYGLASS
lately invented by him,
In the surface of the Moon, in innumerable
Fixed Stars, in Nebulae, and above all
in FOUR PLANETS
swiftly revolving about Jupiter at
differing distances and periods,
and known to no one before the
Author recently perceived them
and decided that they should
be named
THE MEDICEAN STARS

Venice
1610

Great indeed are the things which in this brief treatise I propose for observation and consideration by all students of nature. I say great, because of the excellence of the subject itself, the entirely unexpected and novel character of these things, and finally because of the instrument by means of which they have been revealed to our senses.

Surely it is a great thing to increase the numerous host of fixed stars previously visible to the unaided vision, adding countless more which have never before been seen, exposing these plainly to the eye in numbers ten times exceeding the old and familiar stars.

It is a very beautiful thing, and most gratifying to the sight, to behold the body of the moon, distant from us almost sixty earthly radii, as if it were no farther away than two such measures—so that its diameter appears almost thirty times larger, its surface nearly nine hundred times, and its volume twenty-seven thousand times as large as when viewed with the naked eye. In this way one may learn with all the certainty of sense evidence that the moon is not robed in a smooth and polished surface but is in fact rough and uneven, covered everywhere, just like the earth's surface, with huge prominences, deep valleys, and chasms.

Again, it seems to me a matter of no small importance to have ended the dispute about the Milky Way by making its nature manifest to the very senses as well as to the intellect. Similarly it will be a pleasant and elegant thing to demonstrate that the nature of those stars which astronomers have previously called "nebulous" is far different from what has been believed hitherto. But what surpasses all wonders by far, and what particularly moves us to seek the attention of all astronomers and philosophers, is the discovery of four wandering stars not known or observed by any man before us. Like Venus and Mercury, which have their own periods about the sun, these have theirs about a certain star that is conspicuous among those already known, which they sometimes precede and sometimes follow, without ever departing from it beyond certain limits. All these facts were discovered and observed by me not many days ago with the aid of a spyglass which I devised, after first being illuminated by divine grace. Perhaps other things, still more remarkable, will in time be discovered by me or by other observers with the aid of such an instrument, the form and construction of which I shall first briefly explain, as well as the occasion of its having been devised. Afterwards I shall relate the story of the observations I have made.

About ten months ago a report reached my ears that a certain Fleming had constructed a spyglass by means of which visible objects, though very distant from the eye of the observer, were distinctly seen as if nearby. Of this truly remarkable effect several experiences were related, to which some persons gave credence while others denied them. A few days later the report was confirmed to me in a letter from a noble Frenchman at Paris, Jacques Badovere, which caused me to apply myself wholeheartedly to inquire into the means by which I might arrive at the invention of a similar instrument. This I did shortly afterwards, my basis being the theory of refraction. First I prepared a tube of lead, at the ends of which I fitted two glass lenses, both plane on one side while on the other side one was spherically convex and the other concave. Then placing my eye near the concave lens I perceived objects satisfactorily large and near, for they appeared three times closer and nine times larger than when seen with the naked eye alone. Next I constructed another one, more accurate, which represented objects as enlarged more than sixty times. Finally, sparing neither labor nor expense, I succeeded in constructing for myself so excellent an instrument that objects seen by means of it appeared nearly one thousand times larger and over thirty times closer than when regarded with our natural vision.

It would be superfluous to enumerate the number and importance of the advantages of such an instrument at sea as well as on land. But forsaking terrestrial observations, I turned to celestial ones, and first I saw the moon from as near at hand as if it were scarcely two terrestrial radii away. After that I observed often with wondering delight both the planets and the fixed stars, and since I saw these latter to be very crowded, I began to seek (and eventually found) a method by which I might measure their distances apart.

Now let us review the observations made during the past two months, once more inviting the attention of all who are eager for true philosophy to the first steps of such important contemplations. Let us speak first of that surface of the moon which faces us. For greater clarity I distinguish two parts of this surface, a lighter and a darker; the lighter part seems to surround and to pervade the whole hemisphere, while the darker part discolors the moon's surface like a kind of cloud, and makes it appear covered with spots. Now those spots which are fairly dark and rather large are plain to everyone and have been seen throughout the ages; these I shall call the "large" or "ancient" spots, distinguishing them from others that are smaller in size but so numerous as to occur all over the lunar surface, and especially the lighter part. The latter spots had never been seen by anyone before me. From observations of these spots repeated many times I have been led to the opinion and conviction that the surface of the moon is not smooth, uniform, and precisely spherical as a great number of philosophers believe it (and the other heavenly bodies) to be, but is uneven, rough, and full of cavities and prominences, being not unlike the face of the earth, relieved by chains of mountains and deep valleys. The things I have seen by which I was enabled to draw this conclusion are as follows.

On the fourth or fifth day after new moon, when the moon is seen with brilliant horns, the boundary which divides the dark part from the light does not extend uniformly in an oval line as would happen on a perfectly spherical solid, but traces out an uneven, rough, and very wavy line. . . . Indeed, many luminous excrescences extend beyond the boundary into the darker portion, while on the other hand some dark patches invade the illuminated part. Moreover a great quantity of small blackish spots, entirely separated from the dark region, are scattered almost all over the area illuminated by the sun with the exception only of that part which is occupied by the large and ancient spots. Let us note, however, that the said small spots always agree in having their blackened parts directed toward the sun, while on the side opposite the sun they are crowned with bright contours, like shining summits. There is a similar sight on earth about sunrise, when we behold the valleys not yet flooded with light though the mountains surrounding them are already ablaze with glowing splendor on the side opposite the sun. And just as the shadows in the hollows on earth diminish in size as the sun rises higher, so these spots on the moon lose their blackness as the illuminated region grows larger and larger.

Again, not only are the boundaries of shadow and light in the moon seen to be uneven and wavy, but still more astonishingly many bright points appear within the darkened portion of the moon, completely divided and separated from the illuminated part and at a considerable distance from it. After a time these gradually increase in size and brightness, and an hour or two later they become joined with the rest of the lighted part which has now increased in size. Meanwhile more and more peaks shoot up as if sprouting now here, now there, lighting up within the shadowed portion; these become larger, and finally they too are united with that same luminous surface which extends ever further. And on the earth, before the rising of the sun, are not the highest peaks of the mountains illuminated by the sun's rays while the plains remain in shadow? Does not the light go on spreading while the larger central parts of those mountains are becoming illuminated? And when the sun has finally risen, does not the illumination of plains and hills finally become one? But on the moon the variety of elevations and depressions appears to surpass in every way the roughness of the terrestrial surface. . . .

Let these few remarks suffice us here concerning this matter, which will be more fully treated in our *System of the World.* In that book, by a multitude of arguments and experiences, the solar reflection from the earth will be shown to be quite real—against those who argue that the earth must be excluded from the dancing whirl of stars for the specific reason that it is devoid of motion and of light. We shall prove the earth to be a wandering body surpassing the moon in splendor, and not the sink of all dull refuse of the universe; this we shall support by an infinitude of arguments drawn from nature.

We have now briefly recounted the observations made thus far with regard to the moon. . . . There remains the matter which in my opinion deserves to be considered the most important of all—the disclosure of four PLANETS never seen from the creation of the world up to our own time, together with the occasion of my having discovered and studied them, their arrangements, and the observations made of their movements and alterations during the past two months. I invite all astronomers to apply themselves to examine them and determine their periodic times, something which has so far been quite impossible to complete, owing to the shortness of the time. Once more, however, warning is given that it will be necessary to have a very accurate telescope such as we have described at the beginning of this discourse.

On the seventh day of January in this present year 1610, at the first hour of night, when I was viewing the heavenly bodies with a telescope, Jupiter presented itself to me; and because I had prepared a very excellent instrument for myself, I perceived (as I had not before, on account of the weakness of my previous instrument) that beside the planet there were three starlets, small indeed, but very bright.

Though I believed them to be among the host of fixed stars, they aroused my curiosity somewhat by appearing to lie in an exact straight line parallel to the ecliptic, and by their being more splendid than others of their size. Their arrangement with respect to Jupiter and each other was the following:

East   ✳   ✳   ○   ✳   *West*

that is, there were two stars on the eastern side and one to the west. The most easterly star and the western one appeared larger than the other. I paid no attention to the distances between them and Jupiter, for at the outset I thought them to be fixed stars, as I have said. But returning to the same investigation on January eighth—led by what, I do not know—I found a very different arrangement. The three starlets were now all to the west of Jupiter, closer together, and at equal intervals from one another as shown in the following sketch:

East   ○   ✳   ✳   ✳   *West*

At this time, though I did not yet turn my attention to the way the stars had come together, I began to concern myself with the question how Jupiter could be east of all these stars when on the previous day it had been west of two of them. I commenced to wonder whether Jupiter was not moving eastward at that time, contrary to the computations of the astronomers, and had got in front of them by that motion. Hence it was with great interest that I awaited the next night. But I was disappointed in my hopes, for the sky was then covered with clouds everywhere.

On the tenth of January, however, the stars appeared in this position with respect to Jupiter:

East   ✳   ✳   ○   *West*

that is, there were but two of them, both easterly, the third (as I supposed) being hidden behind Jupiter. As at first, they were in the same straight line with Jupiter and were arranged precisely in the line of the zodiac. Noticing this, and knowing that there was no way in which such alteration could be attributed to Jupiter's motion, yet being certain that these were still the same stars I had observed (in fact no other was to be found along the line of the zodiac for a long way on either side of Jupiter), my perplexity was now transformed into amazement. I was sure that the apparent changes belonged not to Jupiter but to the observed stars, and I resolved to pursue this investigation with greater care and attention.

And thus, on the eleventh of January, I saw the following disposition:

East   ✳ ✳   O   *West*

There were two stars, both to the east, the central one being three times as far from Jupiter as from the one farther east. The latter star was nearly double the size of the former, whereas on the night before they had appeared approximately equal.

I had now decided beyond all question that there existed in the heavens three stars wandering about Jupiter as do Venus and Mercury about the sun, and this became plainer than daylight from observations on similar occasions which followed. Nor were there just three such stars; four wanderers complete their revolutions about Jupiter, and of their alterations as observed more precisely later on we shall give a description here. Also I measured the distances between them by means of the telescope, using the method explained before. Moreover I recorded the times of the observations, especially when more than one was made during the same night—for the revolutions of these planets are so speedily completed that it is usually possible to take even their hourly variations.

Thus on the twelfth of January at the first hour of night I saw the stars arranged in this way:

East   ✳ ∗○ ✳   *West*

The most easterly star was larger than the western one, though both were easily visible and quite bright. Each was about two minutes of arc distant from Jupiter. The third star was invisible at first, but commenced to appear after two hours; it almost touched Jupiter on the east, and was quite small. All were on the same straight line directed along the ecliptic.

On the thirteenth of January four stars were seen by me for the first time, in this situation relative to Jupiter:

East   ∗ ○∗✳∗   *West*

Three were westerly and one was to the east; they formed a straight line except that the middle western star departed slightly toward the north. The eastern star was two minutes of arc away from Jupiter, and the intervals of the rest from one another and from Jupiter were about one minute. All the stars appeared to be of the same magnitude, and though small were very bright, much brighter than fixed stars of the same size.

On the twenty-sixth of February, midway in the first hour of night, there were only two stars:

East   ∗   ○   ✳   *West*

One was to the east, ten minutes from Jupiter; the other to the west, six minutes away. The eastern one was somewhat smaller than the western. But at the fifth hour three stars were seen:

East   ∗   ○ ∗   ✳   *West*

In addition to the two already noticed, a third was discovered to the west near Jupiter; it had at first been hidden behind Jupiter and was now one minute away. The eastern one appeared farther away than before, being eleven minutes from Jupiter.

THE ACHIEVEMENT OF GALILEO

Such are the observations concerning the four Medicean planets recently first discovered by me, and although from these data their periods have not yet been reconstructed in numerical form, it is legitimate at least to put in evidence some facts worthy of note. Above all, since they sometimes follow and sometimes precede Jupiter by the same intervals, and they remain within very limited distances either to east or west of Jupiter, accompanying that planet in both its retrograde and direct movements in a constant manner, no one can doubt that they complete their revolutions about Jupiter and at the same time effect all together a twelve-year period about the center of the universe. That they also revolve in unequal circles is manifestly deduced from the fact that at the greatest elongation from Jupiter it is never possible to see two of these planets in conjunction, whereas in the vicinity of Jupiter they are found united two, three, and sometimes all four together. It is also observed that the revolutions are swifter in those planets which describe smaller circles about Jupiter, since the stars closest to Jupiter are usually seen to the east when on the previous day they appeared to the west, and vice versa, while the planet which traces the largest orbit appears upon accurate observation of its returns to have a semimonthly period.

Here we have a fine and elegant argument for quieting the doubts of those who, while accepting with tranquil mind the revolutions of the planets about the sun in the Copernican system, are mightily disturbed to have the moon alone revolve about the earth and accompany it in an annual rotation about the sun. Some have believed that this structure of the universe should be rejected as impossible. But now we have not just one planet rotating about another while both run through a great orbit around the sun; our own eyes show us four stars which wander around Jupiter as does the moon around the earth, while all together trace out a grand revolution about the sun in the space of twelve years.

from

The Assayer*

In which Galileo ridicules those who prefer the books of men to the great open book of Nature itself.

It seems to me that I detect in Sarsi [an Aristotelian disciple] the fixed persuasion that in philosophizing one has to rely on the opinions of some famous author, as if this mind of ours would remain sterile and barren unless it were wedded to another person's thoughts. Perhaps he thinks philosophy is a book of fiction or some other kind of imaginative work, like the *Iliad* or *Orlando Furioso*, in which it is of least importance that what is written be true.

My dear Sarsi, that's not how the matter stands. Philosophy is written in this grandest of all books which forever lies open before our eyes (I mean the universe), but which cannot be understood if one does not first learn to understand the language and interpret the characters in which it is written. It is written in mathematical language, and the characters are triangles, circles, and other geometrical figures, without which it is humanly impossible to understand a single word; without these, there is only aimless wandering in a dark labyrinth. . . .

I cannot omit to . . . indicate, further, how ill-founded is Sarsi's conclusion that scientific knowledge must be deficient if the number of disciples is few. Perhaps Sarsi thinks that good philosophers are to be found like squadrons of soldiers in every camp. My belief, dear Sarsi, is that they fly like eagles, and not like starlings. Indeed, the truth is that eagles, because they are rare, are rarely seen and even

---

* Translated by Henry Paolucci from Galileo, *Il Saggiatore*.

less often heard; whereas the starlings that fly in throngs, wherever they alight, "filling the sky with shrieks and noises," stir up the whole world. But, if only true philosophers were indeed like eagles, and not rather like the phoenix! My dear Sarsi, "infinite is the throng of fools," that is to say, those who know nothing, myriads there are who know next to nothing of philosophy; some few know a trivial bit of it; fewest of all know a part or two; God alone is He who knows all.

So that, to state exactly the point I want to make regarding the scientific knowledge men can attain by demonstrations and human speculation, I am convinced that as it attains greater perfection, it will propose to draw ever fewer conclusions; fewer still will it pretend to prove, and, as a consequence, so much less will it please, and less, proportionately, will the number of its adherents be. On the other hand, high-sounding titles, coupled with grandiose and numerous promises, attracting the natural curiosity of men, keeping them perpetually tangled up in fallacies and chimeras, never enabling them to taste the sharpness of a single proof, from which the reawakened taste would learn how insipid indeed has been its usual fare—such is the sort of thing that will keep an infinitude of persons busily interested. And it will be the happiest of accidents for anyone, guided by extraordinary natural insight, to be able to free himself from the benighted and confused labyrinths in which he, with everyone else, would otherwise forever wander, ever entangling himself the more.

To judge anyone's views in matters of science, therefore, from the number of followers, I hold to be quite unsound. But, while I believe that the best philosophy will have very few adherents, I do not, as a consequence, maintain conversely that those views and doctrines are necessarily perfect which have the fewest adherents. For I can well conceive of somone entertaining views so erroneous that he would of necessity be forsaken by everyone else.

from the

Dialogue on the Great World Systems*

> *In which, through the lively discourse*
> *of three personages partly drawn from*
> *real life—Salviati, Sagredo and the Aris-*
> *totelian disciple Simplicio—Galileo drami-*
> *tizes his polemic against blind adherence*
> *to authority in the sphere of empirical*
> *knowledge.*

SIMPLICIO: I must confess that I have been ruminating all night of what passed yesterday, and, to say the truth, I have met with many acute, new, and plausible notions; yet I remain convinced by the authority of so many great writers and in particular. . . . I see you shaking your head, Sagredo, and grinning to yourself, as if I had uttered some great absurdity.

SAGREDO: I not only grin but actually am ready to burst with holding myself from laughing outright, for you have put me in mind of a very pretty episode that I witnessed not many years since, together with some others of my worthy friends which I could name to you.

SALVIATI: It would be well that you told us what it was, so Simplicio may not still think that he is the point of your laughter.

SAGR.: Very well. One day at his home in Venice, I found a famous physician to whom some flocked for their studies, while others sometimes came thither out of curiosity to see certain bodies dissected by the hand of a no less learned than careful and experienced anatomist. It chanced upon that day, when I was there, that he was in search of the origin and stem of the nerves, about which there is a famous controversy between the Galenists and Peripatetics. The anatomist shewed how the great trunk of nerves, departing from the brain, their root, passed by the nape of the neck, extended themselves afterwards along the backbone, and branched out through all the body, while only a very small filament, as fine as a thread, went to the heart. Then he turned to a gentleman whom he knew to be a Peripatetic philosopher, and for whose sake he had uncovered and proved everything, and asked if he was satisfied and persuaded that the origin of the nerves was in the brain and not in the heart. The philosopher, after he had stood musing a while, answered: "You have made me see this business so plainly and sensibly that did not the text of Aristotle assert the contrary, which positively affirms the nerves to proceed from the heart, I should be constrained to confess your opinions to be true."

SIMP.: I would have you know, my Masters, that this controversy about the origin of the nerves is not yet so proved and decided, as some may perhaps persuade themselves.

SAGR.: Nor doubtless shall it ever be if it finds such contradictors; but what you say does not at all lessen the extravagance of the answer of that Peripatetic, who against such sensible experience did not produce other experiments or reasons of Aristotle but his bare authority and pure *ipse dixit*.

SIMP.: Aristotle would not have gained so great authority but for the force of his demonstrations and the profundity of his argument. It is requisite that we understand him, and not only understand him, but have so great a familiarity with his books, that we form a perfect idea of them in our minds, so that every saying of his may be always, as it were, present in our memory. He did not write for the vulgar, nor is he obliged to spin out his syllogisms in the trivial ordered meth-

od; on the contrary, using the perturbed order, he has sometimes placed the proof of one proposition among texts which
seem to treat of quite another point. Therefore it is necesary
to be master of all that vast system and to learn how to
connect this passage with that and to combine this text with
another far remote from it; he who has thus studied him will
know how to gather from his books the demonstrations of
every knowable deduction, because they contain all things.

SAGR.: But, good Simplicio, this reaching the desired conclusion by connecting several small extracts which you and
other egregious philosophers easily find scattered throughout the texts of Aristotle I could do as well by the verses
of Virgil or Ovid, composing patchworks of passages which
explain all the affairs of men and secrets of Nature. But
why do I talk of Virgil or any other poet? I have a little
book much shorter than Aristotle or Ovid, in which are
contained all the sciences, and with very little study one
may gather out of it a most perfect system, and this is the
alphabet. For there is no doubt but that he who knows
how to couple and correctly dispose this and that vowel
with the right consonants may gather thence the infallible
answers to all doubts and deduce from them the principles
of all sciences and arts. In the same manner the painter, from
many simple colours laid individually upon his palette, proceeds, by mixing a little of this and a little of that with a
little of a third, to represent lifelike men, plants, buildings,
birds, fishes, or, in a word, counterfeiting whatever object
is visible, though there be not on the palette, all the while,
either eyes, feathers, fins, leaves, or stones. Even more
necessary is that none of the things to be imitated, or any
part of them be actually among colours, if you would be
able therewith to represent all things; for should there be
among them, say, feathers, these would serve to represent
nothing save birds and plumed creatures.

SALV.: And there are certain gentlemen yet living and in
health who were present when a professor in a famous

Academy, hearing the description of the Telescope, said that the invention was taken from Aristotle, though he had not yet seen it. Having his works brought, he turned to a place where the philosopher gives the reason why, from the bottom of a very deep well, one may see the stars in heaven at noonday. Then addressing himself to the company: "See here," says he, "the well, which represents the tube, see here the gross vapours, from whence is taken the invention of the crystals, and see here lastly the sight fortified by the passage of the rays through a diaphanous but more dense and obscure medium."

SAGR.: This is a way to comprehend all things knowable, much like that whereby a piece of marble is one or even a thousand very beautiful statues; but the difficulty lies in being able to discover them. Or we may say that it is like the prophecies of Abbot Joachim or the answers of the heathen oracles, which are not to be understood till after the things foretold have come to pass.

SALV.: And why do you not add the predictions of the astrologers which are seen with like clearness after the event in their horoscopes?

SAGR.: In this manner the alchemists, being led by their melancholy humour, find that all the sublimest minds of the world have written of nothing else in reality than of the way to make gold; but, that they might transmit the secret to posterity without revealing it to the vulgar, they contrived in various ways to conceal the process under several masks. It would make one merry to hear their comments on the ancient poets, gleaning important mysteries hidden under their fables, of the signification of the loves of the Moon, her descending to the Earth for Endymion; her displeasure against Actaeon; and what was meant by Jupiter's turning himself now into a shower of gold and now into flames of fire; and what great secrets of art are contained in that Mercury the Interpreter, in those abductions of Pluto, in those Golden Boughs.

SIMP.: I believe, and in part I know, that in the world there are very extravagant minds whose vanities ought not to redound to the prejudice of Aristotle, of whom I think you speak sometimes with too little respect. Were it only for his antiquity and the great name that he has acquired in the opinions of so many famous men, they should render him honourable with all that profess themselves learned.

SALV.: You do not state the matter rightly, Simplicio. It is those pusillanimous followers of his who would give us cause to think less of him, should we consent to applaud their silly arguments. And you, please tell me, are you so simple as not to realize that, had Aristotle been present and heard the doctor who tried to make him author of the Telescope, he would have been much more displeased with him than with those who laughed at the doctor and his comments? Do you question whether Aristotle, had he but seen the new discoveries in heaven, would not have changed his opinions, amended his books, and embraced the more sensible doctrine, rejecting those silly gulls who go about so timidly to defend whatever he has said? Do those defenders consider that, if Aristotle were such a one as they fancy him to themselves, he would be a man of an untractable wit, an obstinate mind, a barbarous soul, a stubborn will, who, accounting all other men as silly sheep, would have his oracles preferred before the senses, before experience, and before Nature herself? It is the sectators of Aristotle that have given him this authority and not he who has usurped or taken it upon himself. All because it is easier for a man to sulk under another's shield than to shew himself openly, they tremble and are afraid to stir one step from him; rather than admit some alterations in the heaven of Aristotle, they will impertinently deny those they behold in the heaven of Nature.

SAGR.: This kind of people puts me in mind of that sculptor who artfully made a great piece of marble to the image of Hercules, or a thundering Jupiter, I know not which, with such a vivacity and threatening fury that it moved terror in as many as beheld it. Soon he also began to be afraid of it, though all its threatening life was his own workmanship; and his affright was such that he had no longer the courage to affront it with his chisels and mallet.

SALV.: I have many times wondered how these pedantic maintainers of whatever came from Aristotle's pen are not aware how great a prejudice they are to his reputation and credit and how, the more they go about to increase his authority, the more they diminish it. When I see them obstinate in their attempts to maintain those propositions which are manifestly false, and trying to persuade me that to do so is the part of a philosopher, and that Aristotle himself would do the same, it much discourages me in the belief that he has rightly philosophized about other conclusions, for if I could see them concede and change opinion in a manifest truth, I would be more willing to believe that, where they persist, they may have some solid demonstrations by me not understood or even heard of.

SAGR.: Or supposing they were made to see that they have hazarded too much of their own and Aristotle's reputations in confessing that they had not understood this or that conclusion found out by some other man; would it not be a less evil for them to seek for it among his texts, by throwing many of them together, according to the art intimated to us by Simplicio? If his works do contain all things knowable, it must also follow that they contain that too.

SALV.: Good Sagredo, make no jest of this device, which I think you rehearse in too ironical a way. It is not long since a very eminent philosopher composed a book *On the Soul,* wherein, citing the opinion of Aristotle about the soul's being or not being immortal, he alleged many texts (not any of those heretofore quoted by Alexander of Aphrodisias, for in those, he said, Aristotle had not so much as treated of that matter, but others found by himself in more abstruse places) which tended to a pernicious sense. Being advised that he would find it a hard matter to get a license from the Inquisitors, he wrote back to his friend to procure it with all expedition, for, if there was no other difficulty, he would not much scruple to change the doctrine of Aristotle and, with different expositions and different texts, to maintain the contrary opinion, which yet should also be agreeable to the sense of Aristotle.

SAGR.: O most profound Doctor, this! that can command me; for he will not be led around by Aristotle but will lead him by the nose and make him speak as he pleases! See how important it is to learn to seize an opportunity. Nor is it seasonable to have to do with Hercules while he is enraged and beside himself but when he is telling merry tales among the Maeonian damosels. Ah, unheard-of sordidness of servile minds! to make themselves willing slaves, to receive as inviolable decrees, to engage themselves to seem satisfied and convinced by arguments of such efficacy and so manifestly conclusive that they themselves cannot resolve whether they were written to that purpose or serve to prove the assumption in hand! But a greater madness is that they are at variance among themselves, whether the author has held the affirmative part or the negative. What is this but to make an oracle of a wooden image and to run to that for answers, to fear it, to reverence and adore it?

SIMP.: But in case we should give up Aristotle, who is to be our guide in philosophy? Name you some author.

SALV.: We need a guide in unknown and uncouth parts, but in clear thoroughfares, and in open plains, only the blind stand in need of a leader; and, for such, it is better that they stay at home. But he who has eyes in his head and in his mind has to use those for his guide. Yet mistake me not, thinking that I speak this because I am against hearing Aristotle; for, on the contrary, I commend the reading and diligent study of him and only blame the servilely giving one's self up a slave to him, so as blindly to subscribe to whatever he delivers, and receive it for an inviolable decree without search of any further reason. This is an abuse that carries with it the other extreme disorder that people will no longer take pains to understand the validity of his demonstrations. And what is more shameful in public disputes than, while someone is treating of demonstrable conclusions, to have someone else come up with a passage of Aristotle, quite often irrelevant, and with that stop the mouth of his opponent? But, if you will continue to study in this manner, I would have you lay aside the name of philosophers and call yourselves either historians or doctors of memory, for it is not fit that those who never philosophize should usurp the honourable title of philosophers. But it is best for us to return to shore and not launch out further into a boundless gulf, out of which we should not be able to get before night. Therefore, Simplicio, come with arguments and demonstrations of your own, or of Aristotle, but bring us no more texts and naked authorities, for our disputes are about the sensible world and not a paper one.

The Copernican Controversy

THE CONTROVERSY precipitated by Galileo's witty and polemical defense of the "new astronomy" of Copernicus against the entrenched adherents of the "old astronomy" of Ptolemy involved him gradually in further controversy with philosophers and theologians of his time who at first welcomed his work as a major contribution to the perfection of mathematico-physical science. These philosophers and theologians favored neither the Ptolemaic nor the Copernican system, both of which they considered to belong to a lower level of science—the abstract mathematico-physical level—than true philosophy and theology. The selections of this chapter illustrate the most important phases in the development of the complex controversy.

The first reading is the famous letter of Cardinal Bellarmine written in 1615 to the Copernican Foscarini, a friend of Galileo, reminding him, with explicit reference to Galileo, of the traditional distinctions between the kind of knowledge attainable by the application of mathematical principles to the study of empirical phenomena and metaphysical knowledge of the reality underlying phenomena. Galileo, as the next reading, drawn from his notes, makes plain, did not accept the traditional distinction: he was convinced that the application of mathematics to the study of natural phenomena was the surest way of attaining truth regarding the "underlying reality." His attitude of confidence in the certainty of mathematico-physical knowledge provoked many vicious attacks, at first from rival mathematical astronomers, but soon enough from the professional philosophers and theologians. But it also provoked some eloquent defenses, the most noteworthy of which was that written in 1622 by the celebrated Renaissance philosopher Thomas Campanella, from which the third reading in this chapter is drawn.

The heat of the controversy that raged alarmed the censors of the Church whose approval Galileo was anxious to obtain. Galileo had a private meeting on the subject with the Cardinal Barberini, who later became Pope Urban VIII, in which the learned Cardinal, after having admitted that close observation of celestial phenomena accorded with the Galilean-Copernican conception, asked the Tuscan scientist whether it might not be possible for the same phenomena to be "saved" just as well by supposing an actual state of things other than that supposed by the Copernicans. The future Pope then insisted that, before the Copernican system could be accepted as true, in the Galilean sense, it would be necessary to demonstrate not only that it accounted for all the phenomena, but also "that the whole thing cannot, without involving contradiction, be accounted for by any system other than the one you have conceived." So that Galileo might avoid at once quarrels with philosophers and the censures of theologians, the Cardinal Barberini advised him to present his Copernican views as mathematically convenient hypotheses rather than as established truths about "reality."

When Galileo actually undertook to write his *Dialogue on the Great World Systems* (1632) he pretended to take the advice that had been given him. But his pretense was transparent. In the concluding pages of that great work (see the close of the fourth reading in this chapter, p. 109) Galileo allowed the Aristotelian spokesman of his dialogue to express the very thought urged by Cardinal Barberini; only to dismiss it at once with a witticism. For Galileo held dogmatically that the Copernican system was the only true system, true absolutely and without reservations for the future. God knows infinitely more than we can ever know, said Galileo, but what we know mathematically we know as well as He does.

The censors of the Church, when they realized that Galileo had obviously intended his accommodation to the Cardinal's advice to be a transparent pretense, took action against him. To avoid excommunication, and all that it entailed, Galileo was constrained to pronounce the abjuration which is included as the final reading in this chapter.

The Letter of Cardinal Bellarmine to Foscarini*

"I have gladly read the letter in Italian and the essay in Latin that Your Reverence has sent me, and I thank you for both, confessing that they are filled with ingenuity and learning. But since you ask for my opinion, I shall give it to you briefly, as you have little time for reading and I for writing.

"First. I say that it appears to me that Your Reverence and Sig. Galileo did prudently to content yourselves with speaking hypothetically and not positively, as I have always believed Copernicus did. For to say that assuming the earth moves and the sun stands still saves all the appearances better than eccentrics and epicycles is to speak well. This has no danger in it, and it suffices for mathematicians. But to wish to affirm that the sun is really fixed in the center of the heavens and merely turns upon itself without traveling from east to west, and that the earth is situated in the third sphere and revolves very swiftly around the sun, is a very dangerous thing, not only by irritating all the theologians and scholastic philosophers, but also by injuring our holy faith and making the sacred Scripture false. For Your Reverence has indeed demonstrated many ways of expounding the Bible, but you have not applied them specifically, and doubtless you would have had a great deal of difficulty if you had tried to explain all the passages that you yourself have cited.

"Second. I say that, as you know, the Council [of Trent] would prohibit expounding the Bible contrary to the com-

* From *Discoveries and Opinions of Galileo*, translated and edited by Stillman Drake. Copyright © 1957 by Stillman Drake, reprinted by permission of Doubleday & Company, Inc.

mon agreement of the holy Fathers. And if Your Reverence would read not only all their works but the commentaries of modern writers on Genesis, Psalms, Ecclesiastes, and Joshua, you would find that all agree in expounding literally that the sun is in the heavens and travels swiftly around the earth, while the earth is far from the heavens and remains motionless in the center of the world. Now consider whether, in all prudence, the Church could support the giving to Scripture of a sense contrary to the holy Fathers and all the Greek and Latin expositors. Nor may it be replied that this is not a matter of faith, since if it is not so with regard to the subject matter, it is with regard to those who have spoken. Thus that man would be just as much a heretic who denied that Abraham had two sons and Jacob twelve, as one who denied the virgin birth of Christ, for both are declared by the Holy Ghost through the mouths of the prophets and apostles.

"Third. I say that if there were a true demonstration that the sun was in the center of the universe and the earth in the third sphere, and that the sun did not go around the earth but the earth went around the sun, then it would be necessary to use careful consideration in explaining the Scriptures that seemed contrary, and we should rather have to say that we do not understand them than to say that something is false which had been proven. But I do not think there is any such demonstration, since none has been shown to me. To demonstrate that the appearances are saved by assuming the sun at the center and the earth in the heavens is not the same thing as to demonstrate that in fact the sun is in the center and the earth in the heavens. I believe that the first demonstration may exist, but I have very grave doubts about the second; and in case of doubt one may not abandon the Holy Scriptures as expounded by the holy Fathers. I add that the words *The sun also riseth, and the sun goeth down, and hasteth to the place where he ariseth* were written by Solomon, who not only spoke by divine inspiration, but was a man wise above all others, and learned in the human sciences and in the knowledge of all created things, which wisdom he had from God; so it is not very likely that he would affirm something that was contrary to demonstrated truth, or truth that might be demonstrated. And if you tell me that Solomon spoke according to the appearances, and that it seems to us that the sun goes round when the earth turns, as it seems to one aboard ship that the beach moves away, I shall answer thus. Anyone who departs from the beach, though to him it appears that the beach moves away, yet knows that this is an error and corrects it, seeing clearly that the ship moves and not the beach; but as to the sun and earth, no sage has needed to correct the error, since he clearly experiences that the earth stands still and that his eye is not deceived when it judges the sun to move, just as he is likewise not deceived when it judges that the moon and the stars move. And that is enough for the present."

from

Notes Indicating Galileo's Views on the Arguments of Cardinal Bellarmine's Letter*

"One reads on the verso of the title page of Copernicus' book a certain preface to the reader which is not by the author, as it speaks of him in the third person and is unsigned. There it is blandly stated that Copernicus did not believe his system to be true at all, but only claimed to advance it for the calculation of heavenly motions, and finished his reasoning by concluding that it would be foolish to take his theory as real and true. This conclusion is so positively stated that anyone who did not read further, and thought this to have been put there with the author's consent, might well be excused for his mistake. But what value can we place on the opinion of a person who would judge a book by reading no more than a brief preface of the printer and bookseller? I leave this to everyone to judge for himself; and I say that this preface can be nothing but a word from the bookseller to assist the vending of the work, which would have been considered a monstrous chimera by people in general if it had not been qualified in some such way—and generally the buyer reads no more than such a preface before purchasing a book. And that this preface was not only not written by the author, but that it was placed there without his knowledge, to say nothing of his consent, is made manifest by the misuse of certain terms in it which the author would never have permitted."

[Elsewhere in these notes Galileo makes this point-by-point reply to Cardinal Bellarmine's written opinion]:

* From *Discoveries and Opinions of Galileo*, translated and edited by Stillman Drake. Copyright © 1957 by Stillman Drake, reprinted by permission of Doubleday & Company, Inc.

"1. Copernicus assumes eccentrics and epicycles; not these, but other absurdities, were his reason for rejecting the Ptolemaic system.

"2. As to philosophers, if they are true philosophers (that is, lovers of truth), they should not be irritated; but, finding out that they have been mistaken, they must thank whoever shows them the truth. And if their opinion is able to stand up, they will have cause to be proud and not angry. Nor should theologians be irritated; for finding such an opinion false, they might freely prohibit it, or discovering it to be true they should be glad that others have opened the road to the discovery of the true sense of the Bible, and have kept them from rushing into a grave predicament by condemning a true proposition.

"As to rendering the Bible false, that is not and never will be the intention of Catholic astronomers such as I am; rather, our opinion is that the Scriptures accord perfectly with demonstrated physical truth. But let those theologians who are not astronomers guard against rendering the Scriptures false by trying to interpret against them propositions which may be true and might be proved so.

"3. It may be that we will have difficulties in expounding the Scriptures, and so on; but this is through our ignorance, and not because there really are, or can be, insuperable difficulties in bringing them into accordance with demonstrated truth.

"4. . . . It is much more a matter of faith to believe that Abraham had sons than that the earth moves. . . . For since there have always been men who have had two sons, or four, or six, or none . . . there would be no reason for the Bible to affirm in such matter anything contrary to truth. . . . But this is not so with the mobility of the earth, that being a proposition far beyond the comprehension of the common people. . . .

"5. As to placing the sun in the sky and the earth outside it, as the Scriptures seem to affirm, etc., this truly seems to me to be simply . . . speaking according to common sense; for really everything surrounded by the sky is in the sky. . . .

"6. Not to believe that a proof of the earth's motion exists until one has been shown is very prudent, nor do we demand that anyone believe such a thing without proof. Indeed, we seek, for the good of the holy Church, that everything the followers of this doctrine can set forth be examined with the greatest rigor, and that nothing be admitted unless it far outweighs the rival arguments. If these men are only ninety per cent right, then they are defeated; but when nearly everything the philosophers and astronomers say on the other side is proved to be quite false, and all of it inconsequential, then this side should not be deprecated or called paradoxical simply because it cannot be completely proved. . . .

"7. It is true that to prove that the appearances may be saved with the motion of the earth . . . is not the same as to prove this theory true in nature; but it is equally true, or even more so, that the commonly accepted system cannot give reasons for those appearances. That system is undoubtedly false, just as . . . this one may be true. And no greater truth may or should be sought in a theory than that it corresponds with all the particular appearances.

"8. No one asks that in case of doubt the teachings of the Fathers be abandoned, but only that the attempt be made to gain certainty in the matter questioned. . . .

"9. We believe that Solomon and Moses and all the other holy writers knew the constitution of the universe perfectly well, as they also knew that God did not have hands or feet or wrath or prevarication or regret. We cast no doubt on this, but we say that . . . the Holy Ghost spoke thus for the reasons set forth.

"10. The mistake about the apparent motion of the beach and stability of the ship is known to us after we have frequently stood on the beach and observed the motion of the boat, as well as in the boat to observe the beach. And if

we could stand thus now on the earth and again on the sun
or some other star, we might gain positive and sensory
knowledge as to which moved. Yet looking only from these
two bodies, it would always appear that the one we were
on stood still, just as to a man who saw only the boat and
the water, the water would always seem to run and the
boat to stand still. . . . It would be better to compare two
ships, of which the one we are on will absolutely seem to
stand still whenever we can make no other comparison than
between the two ships. . . .

"Besides, neither Copernicus nor his followers make use
of this appearance of the beach and the ship to prove that
the earth moves and the sun stands still. They use it only
as an example that serves to show . . . the lack of contradiction between the simple sense-appearance of a stable
earth and a moving sun if the reverse were really true. For
if nothing better than this were Copernicus' proof, I believe no one would endorse him."

from

The Defense of Galileo*

FOREWORD

by THOMAS CAMPANELLA

It is now essential to discuss two questions: first, should the new philosophy be permitted to search for truth, and secondly, is it desirable or allowable to suppress both the Aristotelian sect and the authority of heathen philosophers, and for our Christian schools to teach the new philosophy in harmony with Sacred Doctrine? A third fundamental controversy has been precipitated by those who condemn the exalted philosophy of Galileo the Florentine because they believe it opposed to the dogmas of Holy Scripture. I shall reply as the truth appears to me.

. . . .

I ask therefore: Is the philosophy which Galileo has made famous and important in harmony with or opposed to the Holy Scriptures? Five chapters will complete this inquiry. In the first I shall set forth the arguments and contentions which oppose Galileo; in the second, those which support him. I shall develop in the following chapter three hypotheses which lead to a double conclusion, and answer in the fourth the arguments brought against Galileo. In the fifth and last I shall discuss [and in part refute] the arguments which favor him.

* From Thomas Campanella, *Defense of Galileo*, translated by Grant McColley, Smith College *Studies in History*, Vol. XXII, Nos. 3-4 (April-July 1937).

from

CHAPTER II

THE ARGUMENTS WHICH SUPPORT GALILEO

1. I mention first that *De Revolutionibus Orbium Caelestium Libri VI*, which Copernicus grounded upon observations of the heavens made in 1525, was presented for publication by respected and reputable theologians. By this act they expressed a judgment that the book contains nothing inimical to Catholic faith, despite the fact that it argues for the motion of the earth, the immobility of the Firmament, and places the Sun in the center of the starry heaven. Galileo has made no additions to the system advanced by Copernicus, and if the *Revolutions* does not adversely affect Catholic faith, neither does the work of Galileo.

2. Pope Paul III, to whom Copernicus dedicated the *Revolutions,* approved the book. Likewise did the Cardinal [Schonberg] who, as the prefatory epistle shows, begged to have the unprinted manuscript transcribed at his own expense. At the time of Paul III distinguished genius flourished within the Church, and the Pontiff summoned from afar men of ability, spirit, and noble mind, and supported and adorned them with honor. It indeed would be strange if such able men had imitated blind moles by rejecting Copernicus. Our contemporaries who oppose his theory are not so great in reputation. I add that Galileo, who seeks always more precise observations, has [with the telescope] sharper eyes than Argus.

3. After Copernicus had published the *Revolutions,* Erasmus Reinhold, John Stadius, Michael Maestlin, Christopher Rothmann and many others accepted his theory. Later astronomers thought themselves unable to prepare accurate ephemerides unless they employed the calculations of Co-

pernicus. They also found themselves wholly at a loss to explain celestial motions without violation of mathematical principles and in harmony with the judgment of reason and of all people, unless they employed the Copernican hypothesis. Nor is the heliocentric theory as recent as Copernicus. As the result of his observations of celestial phenomena, Francesco [Domenico] Maria of Ferrara previously had taught that a new system of astronomy should be construed. His disciple Copernicus built this system.

4. The learned Cardinal Cusanus accepted the hypothesis, and acknowledged that other suns and planets move in orbits in the starry heaven. Nolanus [Bruno], whose errors cannot be named, also supported the theory, as did other philosophers. (Nor were the heretics so much condemned for upholding the hypothesis; rather, Catholics were forbidden to read their books.) Outstanding among Copernican advocates is John Kepler, mathematician to the Emperor, who defends the theory in his prefatory dissertation to the *Starry Messenger* of Galileo. I also may add to them William Gilbert, in his *On the Magnet,* and innumerable other Englishmen. Giovanni Antonio Magini, mathematician of Padua, declared in his *Ephemerides* from 1581 to the present year of 1616 that he accepted the calculations and positions set forth by Copernicus and Reinhold. He boldly criticized other cosmological theories in many of his letters.

5. For more than thirty years the Jesuit and Aristotelian, Reverend Father Clavius, supported the Ptolemaic hypothesis. When he observed, however, that the planets Mercury and Venus revolve about the Sun, he advised astronomers in the final edition of his *Works* that some system other than the Ptolemaic must be sought. One recent astronomer, Fictus Apelles, gave serious consideration to this admonition, and as a result of both the warning and his observation of spots on the Sun, looked with favor upon the hypothesis of Galileo and Copernicus.

6. Galileo's theory of the motion of the earth, of a central

Sun, and of the systems of stars with waters and earthy elements is indeed an ancient conception. It comes from the mouth of Moses himself, and from Pythagoras. Saint Ambrose tells us that although Pythagoras was born in a Grecian city, he was a Jew with Moses. He taught disciples in Greece, and after going into Italy, in Crotona of Calabria. His beliefs [later] were attacked by Aristotle, not with mathematical demonstrations, but with insane reasons and moralistic, rustic conjectures. According to the statements of Giovanni Pico della Mirandola and Saint Ambrose, Aristotle condemned Pythagoras in the same fashion as he condemned the *Books of Moses*. He was unable by syllogistic logic to understand their divine mystery, recondite thought, and sublime majesty. Such an attack upon Moses vindicates our great Galileo from similar insults cast by the Greeks.

As not only Ovid, but as many historians declare, Numa Pompilius, disciple of Pythagoras and the wisest emperor of Rome, supported the hypothesis. Pliny truthfully states, despite the denial of some writers, that having been commanded by the oracle of Delphos to erect and dedicate a statue to the wisest of the Greeks, the Roman Senate dedicated a statue to Pythagoras and named him the most learned of the philosophers. Those who blacken the doctrine and scientific method of Galileo defame Moses, Rome, and Italy. They set Aristotle above Pythagoras at a time when long-buried truth is gleaming forth. But this is not our greatest sin; this sin is that we have neither made known the new earths, the new systems in the heavens, and the new celestial phenomena, nor have declared abroad the harmony of Scripture and Copernican philosophy.

7. From the time of Caselli and Francesco Maria of Ferrara until the present, theologians did not condemn Copernican astronomy. Moreover, reputable men ordered that the *Revolutions* be printed. Such theologians are not fewer today. The traducers of Galileo have fomented insurrection,

not because of their zeal for the teaching of Christ, but from ignorance and jealousy.

8. Because heaven is immobile, Holy Scripture names it the Firmament. The earth therefore must rotate, and the Sun stand in the center of the world. As Copernicus and his followers prove and the followers of Ptolemy now admit, this system explains all phenomena and accords with all principles of mathematics.

9. The spots on the Sun, the new stars in the heavens, and comets above the moon demonstrate clearly that the stars are worlds constructed of material substance.

10. It is not possible to explain correctly and satisfactorily the text of Moses unless the stars are so constructed. We prove this below from the most saintly Fathers.

11. In his *Answers to the Orthodox,* Saint Justin teaches us that when controversy regarding the shape and movement of heaven arose between Christians and heathens, the heathens declared heaven was spherical and mobile, and the Christians that it was a vault which stood immobile. Many Fathers of the Church describe heaven as the Firmament because it is unmoved. . . .

I make a further point in considering the question whether it is expedient to construct a new philosophy or to retain the old. In ancient times the servant philosophy who became haughty to her mistress theology was driven out as was Hagar. The sons of Israel are therefore in part called Jewish and in part Azotic. As Esdras desired, the alien wives were seized and cast out by the sons of Judah; that is, by the doctrines of the saints. As we have shown and Galileo does not cease to show, the sciences will be restored by examining the world, the book of God. Saint Thomas says in the *Summa Theologica,* I, question 1, articles 5 ff., that unbelievers are quoted in theological works as witnesses against themselves, not as judges or as witnesses against us. It is a marvel, and one which amazed Bembo, that the unbelieving heathens not only are received as masters, but as masters of the theologians. I pause with this.

Because knowledge should be Christian, they lack understanding who forbid and prohibit philosophy among the followers of Christ. They are similar to the Emperor Julian, who outlawed from the faith and interdicted all the sciences of the Christians. As a result theology, now destitute of her servants, could not call men to the towers of the city of God. This misfortune Saint Thomas discusses in the *Reply*, and calls them Julianites who demanded that monks be forbidden to read secular books and sought to bar us from study of the book of Christ, which is the world. The Scripture of God provides them no excuse. Truly the words "Be ye unwilling to know more than is proper," and, "Who is seen to be wise for himself is foolish," are not against us. On the contrary, they support us. God does not prohibit philosophic inquiry, but rather inquiry concerning things beyond philosophy, as if we knew all and at our pleasure placed human wisdom above revealed doctrine; or as Gentiles and heretics presumed to circumscribe divine truth and to place the gleaming lamp of Scripture under the Aristotelian bushel. Only because of such presumption does the Book of Job criticize human wisdom, and Isaiah castigate astrologers. It now is established that [within their spheres] wisdom is a divine virtue and astrology a useful science. When Machiavellianism is exalted above divinity, and when man excludes God and considers his proper study an inquiry into what is above nature, human wisdom is properly condemned. Astrology is justly rebuked when, like that ancient astrology which raised itself above the prophets in Babylon, it presumes to predict with certainty events not subject to prophecy, or when by conjecture as to future occurrences, it handicaps a sober analysis of affairs. So I state emphatically regarding the other sciences.

Appendix. It is an essential part of the glory of the Christian religion that we permit [Galileo's] method of discovering new knowledge and of rectifying the old. By so doing we may not be required to cleanse the nails and hair of the heathen. We have begged our philosophy from condemned Gentiles, as if we would make them our superiors. Nevertheless, it is necessary for the glory of our religion that we permit, not the continued insults of Machiavelli and Julian, but that men may observe Christ and the wisdom of God. This point we discussed above, citing Saint Augustine, and more extensively in *Anti-Machiavelli*, where we say that approbation of science by Christianity may prove a great bond among those other bonds which bind us to the Church of God. This I believe. Why now do we break this bond?

I have shown that liberty of thought is more vigorous in Christian than in other nations. Should this be true, whosoever prescribes at his own pleasure bounds and laws for human thought, as if this action were in harmony with the decrees of Holy Scripture, he not only is irrational and harmful, but also is irreligious and impious. I say as much of him who teaches and accepts no interpretation but his own, and subjects Scripture to his beliefs or to those of a second writer. Such a practice exposes Holy Scripture to the jest of philosophers and the derision of unbelievers and heretics. It closes one avenue to faith and calls men from, rather than to the high tower of devotion. It creates infidels and injures the Holy Spirit. Augustine in *The Christian Doctrine*, Saint Chrysostom writing on the Psalms, Ambrose and Origen in all their works, and Gregory in *Morals*, XV, declare that analysis of the varied meanings of Scripture may be most pregnant and fruitful. Effective discussion is made sterile by bondage. As Augustine teaches in I *Of the Trinity*, Saint Thomas in *Summa Theologica*, I, question 1, article 10, and Cardinal Cajetan, Scripture is most fruitful in meaning when interpreted both in the mystical and in the literal sense. Indeed, as Saint Thomas sets forth in I, question 32, article 4, a passage in Holy Writ may be given all interpretations and expositions which do not directly or indirectly contradict other biblical passages.

In the fashion of many commentators, Saint Thomas presents in his *Tract Against the Errors of the Greeks*, X, question 18, the sound and equitable position which Augustine announced in the first chapter of his *Commentary on Genesis:* "To the end that they may be protected from the distressing ridicule of secular writers, the words of the Scriptures are expounded in many ways." He further declared in *Of the Trinity* that this practice is one among the various methods by which the cavils of heretics are avoided. In the preface to his *Tract*, Saint Thomas states: "I first assert that many Scriptural passages do not pertain to dogmas of faith but rather to doctrines of philosophy. It does great violence to such passages to affirm or to deny them as if they are pertinent to these dogmas. Augustine said truly in *Confession*, V, 'When I hear an erroneous opinion set forth by any Christian who is ignorant of what philosophers have determined regarding the heaven and stars, and the motions of Sun and Moon, I look with tolerance upon the man and his opinion. Although he is ignorant of the order and nature of the universe, I do not consider him injured if he believes nothing unworthy of the Lord and Creator of all. He is injured however if he affirms obstinately what he is ignorant of, and considers scientific beliefs a part of the doctrines of faith.'

"That such a conception of scientific opinion is patently prejudicial to our religion, Augustine again states in the initial chapter of his *Commentary on Genesis:* 'It is greatly to be guarded against, and is pernicious and shameful for a Christian to speak of physical phenomena as if he were discussing Scripture. Because of this practice some infidel will declaim foolishly. Regardless of the theory advanced by the Christian, if the phenomena of heaven are observed to depart from the hypothesis, the heathen scarcely will restrain his laughter. Nor is it the great misfortune that human theories should be found incorrect, but that our Christian writers are rebuked as ignorant and are thought to believe erroneous hypotheses by men without the faith, men concerning whose salvation we are troubled because of their utter decay. It therefore appears to me desirable that we understand the spheres of philosophy and of our faith do not conflict, and that theories of philosophy are not to be defended as are the dogmas of faith. New theories always may be introduced under the name of philosophy, and such theories should not be opposed on the grounds that they are contrary to faith, nor should worldly wisdom be given occasion to condemn the doctrines of faith.'" Thus Saint Thomas declares with Augustine. From these statements it is apparent that our contemporaries who support Aristotelianism as a doctrine of faith because Saint Thomas expounded this philosophy, do so ignorantly and in opposition to the Fathers. Indeed, as I shall demonstrate more fully in Chapter IV, Saint Thomas teaches that Aristotelianism is not a doctrine of faith.

Among those who challenge natural philosophy in the name of faith is Ulisse Albergotti, who affirms the Moon shines with her own light because Scripture says, "The Moon shall not give her light." His argument is based upon use of the word *her*, but the passage permits many other interpretations. However, the astonishing thing is that having been correct in their major premise regarding the separate spheres of Scripture and philosophy, Augustine and other Fathers erred in the proposition of their syllogism. Lactantius declared in the *Divine Institutions*, III, 25, and Augustine in XVI *The City of God* that antipodes do not exist, because the men thought to be there could not come from the loins of Adam. Therefore the existence of antipodes contradicts Scripture, which describes all men as created from one. To this argument they add scientific evidence. In the year of our Lord 500, Procopius Gaza built from the writings of all the Fathers a chain of interpretations of Scripture, and proved by them that antipodes do not exist. Because of the Fathers' statements and the authority of Holy Scripture, Saint Ephrem placed the earthly paradise in the opposite hemisphere discovered by Columbus. Indeed, those writers who supposed antipodes were judged heretics by various Fathers. Nevertheless, many navigators have demonstrated the falsity of their opinion. If the existence of antipodes is contrary to the Scripture of God, as these Fathers declared; or if there is an earthly paradise or hell or purgatory, as Dante, Isidore and others have believed, it follows that the truth disclosed by Columbus is discordant with or contrary to Divine Scripture.

In harmony with the philosopher Xenophanes, Procopius and others state that our earth is founded on the waters and floats upon them. They prove this belief from David's lines in Psalm 135, "He founded the earth upon the waters," and Psalm 123, "He established it upon the seas." But the earth now is seen hanging in the middle of the world, sustaining itself and the waters, and not sustained by the seas below. Nor according to nature could the waters be placed underneath the earth unless they were at its center, for in the ordering of nature's system the several parts seek the center, and one and all practice this conservation. The parts of the Sun likewise push toward the center of the Sun and the parts of the Moon to the center of the Moon. What anxiety tormented Saint Ambrose because the movement of heaven could not be explained by rising and falling motion, with the result that with Chrysostom and other Fathers he inclined to belief in its quiescence. But such arguments have little weight in science, and it is wrong and pernicious to set them forth as matters of faith.

Moreover, Bishop Philastrius described certain beliefs as part of the Christian faith which are contrary to it, and was ridiculed by both Catholic and heretic for his assertions that the age of the world is as great as that which he computed, and that when God breathed into Adam the breath of life, he did not give life to Adam but rather the Holy Spirit. Bede was more cautious when he maintained that dropsy may arise from a fault of the bladder. So was Saint Thomas when he stated under the influence of Aristotle and despite Albertus and Avicenna that men could not live on the equator. Because these opinions were not considered matters of faith (as Saint Thomas regarded that of the flaming sword), they were refuted by evidence from geography and medicine without injury to religion. But the theologians erred abominably who taught that the torrid zone of the earth was the flaming sword of the angel guarding the road to paradise, for it has been found that this zone does not impede travellers and navigators. What said the unbelievers and Mohammedans when they heard this notion maintained as a doctrine in harmony with Christian Scripture? We may reply to the censure of the Mohammedans, for they suppose the existence of seven earths under this earth, sustained by cattle and fish, with the head in the east and tail in the west. Yet it is small consolation to publish abroad the faults of others when we err ourselves.

If Galileo shall have demonstrated conclusively the things which he affirms, they shall bring forth among heretics no slight mockery of our Roman theology. Particularly is this true since both his hypothesis and the telescope have been accepted with avidity by many men in Germany, France, England, Poland, Denmark, and Sweden. If the hypothesis of Galileo be false, it will not disturb theological doctrine, for not all that is untrue is contrary to faith in the militant Church, although peradventure it may be in the Church triumphant. Indeed, were all that is false contrary to faith, our discovery of the errors of the Saints in natural philosophy had proved them heretics. Moreover if Galileo's theory be unsound, it will not endure. I believe therefore that his type of philosophy should not be forbidden. We are aware how vigorously the Ultramontanes complained because of the decrees of the Council of Trent. The new philosophy will be embraced eagerly by heretics and we shall be ridiculed. What shall they think when they hear we have rebelled against physics and astronomy? Will they not immediately cry out that we block the way, not only of nature, but of Scripture?

. . . .

I believe with Saint Thomas and Augustine that we cannot spurn sublime genius without bringing ridicule upon Scripture, or raising the strong suspicion that with atheists we believe contrary to its Sacred Word. Bellarmine himself declares that at this time heretics do not endanger Roman theology, and for this reason alone it is unnecessary that the investigations of Galileo should be forbidden and his books suppressed, a misfortune which is about to occur. Our enemies will seize eagerly upon this action, and proclaim it abroad.

. . . .

In the above statements, discussions, and writings I at all times submit myself to the correction and better judgment of our Holy Mother the Roman Church. Farewell most illustrious Cardinal Gaetani, protector of Italian excellence.

from the

Dialogue on the Great World Systems*

> *In which the careful reader will find justification for Stillman Drake's recent statement, cited in the Introduction, on the perennial value of Galileo as a teacher of science: "All later attempts to explain scientific method and define scientific truth, however much more logical and thorough, have been considerably less effective."*

INTERLOCUTORS: *Salviati, Sagredo, Simplicio*

SALVIATI: It was our resolution of yesterday that we should discourse as distinctly and concretely as we could of the natural reasons, and their efficacy, that have been hitherto alleged on one side by the maintainers of the Aristotelian and Ptolemaic positions and on the other by the followers of the Copernican system. Copernicus, placing the Earth among the movable bodies of Heaven, makes of it a globe like a planet; therefore, it would be good if we began our disputation with the examination of what and how great the strength of the Peripatetics' argument is when they demonstrate that this hypothesis is impossible. They say that it is necessary to introduce in Nature different substances, that is, the celestial and the elementary; the first unchangeable and immortal, the other alterable and corruptible. This argument Aristotle handles in his book *On Heaven*, bringing it in first by some discourses dependent on general assump-

* Galileo, *Dialogue on the Great World Systems,* the Salusbury translation revised by Giorgio de Santillana. Copyright 1953 by the University of Chicago Press; reprinted by permission of the University of Chicago Press.

tions and afterwards confirming it by way of sense experience and particular demonstration. Following the same method, I will propound and freely speak my judgment, submitting myself to your censure, and particularly to Simplicio, a stout champion and contender for the Aristotelian doctrine.

The first step of the Peripatetic arguments is that in which Aristotle proves the integrity and perfection of the world, telling us that it is not a simple line or a bare surface but a body adorned with longitude, latitude, and depth; and, because there are no more dimensions than these three, the world, in having them, has all and, having all, is to be concluded perfect.

[Galileo's repudiation of the traditional distinction between a physics of the four elements here below and a physics of the quintessence of the celestial spheres. There is hereafter to be only one physics for all phenomena of the inanimate world.]

. . . leaving the general contemplation of the whole, let us descend to the consideration of the two parts into which Aristotle in his first division separates it, each very different and almost contrary to one another; namely, the celestial and the elementary. The first ungenerated, incorruptible, unalterable, impassible, etc.; and the latter exposed to a continual alteration, mutation, etc. This difference, as from its original cause, he derives from the diversity of local motions, and in this method he proceeds. . . . He comes to assert, as a thing known and manifest, that the motions directly upwards or downwards naturally agree to fire and earth. Therefore, besides these which are near us, there must be in Nature another body, to which the circular motions may agree, which shall be so much more excellent as the circular motion is more perfect than the straight. But how much more perfect one is than the other, he determines from the greatness of the circular line's perfection above the straight line, calling the former perfect and the latter imperfect—imperfect because, if infinite, it wants a termination; and if it be finite, there is yet something beyond which it may be prolonged. This is the basis, groundwork, and master-stone of all the fabric of the Aristotelian world, upon which he constructs all the other properties, of being neither heavy nor light, of being ungenerated, incorruptible, and exempt from all motion save only the local, etc. All these characters he affirms to be proper to a simple body that is moved circularly; and the contrary qualities of gravity, levity, and corruptibility, etc., he assigns to bodies naturally moving in a straight line. . . .

SALV.: This principle then established, one may immediately conclude that, if the integral components should be by their nature movable, it is impossible that their motions should be straight or other than circular. This reason is sufficiently easy and manifest; for whatever moves with a straight motion changes place and, continuing to move, moves by degrees farther away from the term from whence it departed and from all the places through which it has successively passed. If such motion naturally suited with it, then it was not, in the beginning, in its proper place; and so the parts of the world were not disposed with perfect order. But we suppose them to be perfectly ordered; therefore, as such, it is impossible that by nature they should change place and consequently move in a straight motion. Moreover, the straight motion being by nature infinite, because the straight line is infinite and indeterminate, it is impossible that any movable body can have a natural principle of moving in a straight line, namely, toward the place whither it is impossible to arrive, there being no predetermined limit; and Nature, as Aristotle himself well says, never attempts to do that which cannot be done or to move whither it is impossible to arrive. And if anyone should yet object that while the straight line, and consequently the motion in it, can be infinitely prolonged, that is to say, is interminate; and that yet, nevertheless, Nature has, so to speak, arbitrarily assigned some limits, or places, and given natural instincts to its natural bodies to move unto the same; I will reply that this might perhaps be fabled to have come to pass in the first Chaos, where indiscrete matters confusedly and inordinately wandered. Then, to regulate them, Nature very appositely may have made use of the straight motions, by which, in the same way as the well-constituted parts, by moving, disorder themselves, so these, which were disposed chaotically, were ranged in order by this motion. But, after their exquisite distribution and collocation, it is impossible that there should remain natural inclinations in them of moving any longer in a straight motion, from which now would ensue their removal from their proper and natural place; that is to say, their disorder. We may therefore say that the straight motion serves to bring the matter into place so as to erect the work; but, once erected, it has to rest immovable or, if movable, to move only circularly. . . .

I conclude, therefore, that the circular motion is the only natural one consistent with natural bodies, i.e., those parts of the universe which are constituted in an excellent arrangement; and that the rectilinear, at the most that can be said for it, is assigned by Nature to bodies and their parts, at such times as they shall be out of their proper places, constituted in a disorderly disposition, and for that cause needing to be brought back by the shortest way to their natural state. Hence, I think that it may be rationally concluded that, for the maintenance of perfect order among the parts of the Universe, it is necessary to say that bodies are movable only circularly. And, if there be any that do not move circularly, these of necessity are immovables; there is nothing besides rest and circular motion that is suited to the conservation of order. And I wonder not a little myself that Aristotle, who held that the terrestrial globe was placed in the center of the Universe, and there remained immovable, should not say that some natural bodies are movable by Nature and others immovable; especially having before defined Nature to be the principle of Motion and Rest.

SIMP.: Aristotle, though of a very perspicacious wit, would not strain it further than needed: he held in all his arguments that the evidence of the senses was to be preferred before any reasons founded upon strength of wit and said that those who should deny the testimony of sense deserved to be punished with the loss of that sense. Now who is so blind who does not see that the parts of the Earth and water, being heavy, will move naturally downwards, namely, towards the center of the Universe, assigned by Nature herself for the end and term of straight motion downwards. And who likewise does not see that fire and air move right upwards toward the sphere of the lunar orbit, as to the natural end of motion upwards? And this being so manifest, and we being certain that "the whole and the parts have the same reason," why may we not assert, if for an evident proposition, that the natural motion of earth is rectilinear motion *to* the center, and that of fire, straight *from* the center?

SALV.: The most that you can pretend from your discourse, were it granted to be true, is that, from the fact that the parts of the Earth when removed from the whole, namely, from the place where they naturally rest—that is, reduced to a debased and disordered disposition—return to their place spontaneously, and therefore naturally in a straight motion (it being granted that the whole and parts have the same reason), it may be inferred that the terrestrial globe, if it were removed violently from the place assigned it by Nature, would return by a straight line. This, as I have said, is the most that can be granted you, and that only for want of examination; but he who will review these things exactly will first deny that the parts of the Earth, in returning to the whole, move in a straight line, not in a circular or mixed one. Then you would have enough to do to demonstrate the contrary, as you shall plainly see in the answers to the particular reasons and observations alleged by Ptolemy and Aristotle. Secondly, another could say that the parts of the Earth do not move toward the center of the world, but to unite with the whole of it, and this for the reason that they naturally incline towards the center of the terrestrial globe, by which inclination they conspire to form and preserve it. What other All or what other center would you find for the world, to which the whole terrestrial globe, being removed from there, would seek to return, so that the reason of the whole might be like to that of its parts? It may be added that neither Aristotle nor you can ever prove that the Earth is actually in the center of the universe; but, if any center may be assigned to the Universe, we might rather find the Sun placed in it, as you shall understand by the sequel.

. . . none of the conditions whereby Aristotle distinguishes the celestrial bodies from the elementary has any foundation other than what he deduces from the diversity of their natural motions; so that, if it is denied that the circular motion is peculiar to celestial bodies, and affirmed instead that it is agreeable to all naturally movable bodies, one is led by necessary conseqence to say either that the attributes of generated or ungenerated, alterable or unalterable, partable or unpartable, etc., equally and commonly apply to all bodies, as well to the celestial as to the elementary, or that Aristotle has badly and erroneously deduced those from the circular motion which he has assigned to celestial bodies.

SIMP.: This manner of thinking tends to the subversion of all natural philosophy and to the disorder and upsetting of Heaven and Earth and the whole Universe. But I believe the fundamentals of the Peripatetics are such that we need not fear that new sciences can be erected upon their ruins.

SALV.: Take no excessive thought in this place for heaven or Earth and do not fear their subversion or the ruin of philosophy. As to heaven, your fears are in vain, for you hold that unalterable and impassible; as for Earth, we strive to ennoble and perfect it, for we try to make it like the celestial bodies and, as it were, place it in Heaven, whence your philosophers have exiled it. Philosophy itself cannot but receive benefit from our disputes, for, if our concepts prove true, new discoveries will have been made; if false, the original doctrine will be more confirmed. Rather bestow your care on philosophers and help and defend them; as for the science itself, it cannot but improve.

And now, to return to our purpose, be pleased to produce freely that which presents itself to you in confirmation of that great difference which Aristotle puts between the celestial bodies and the elementary parts of the world when he makes the former ingenerable, incorruptible, unalterable, etc., and the latter corruptible, alterable, etc.

SIMP.: I do not yet see that Aristotle has any need of help, standing as he does stoutly and strongly on his feet, scarcely being assaulted, much less foiled, by you. And what shield will you choose against his first assault? Aristotle writes that whatever is generated is made out of a contrary in some subject and likewise is corrupted in some certain subject from a contrary into a contrary, so that corruption and generation occur only in contraries; but the movements of contraries are contrary. If, therefore, no contrary can be assigned to a celestial body, because no motion is contrary to a circular motion, then Nature has done very well to exempt that body from contraries, which was to be ungenerated and incorruptible. This fundamental first confirmed, it follows immediately that it is inaugmentable, unalterable, impassible, and finally eternal and a suitable habitation to the immortal deities, conformable to the opinion of all men that have any idea of the gods. He afterwards confirms the same by sense: for, as witness all times past, according to memory or tradition, nothing has been removed from the whole outward heaven or any of its proper parts. Next, as to circular motion, Aristotle proves in many ways that no other is contrary to it; without reciting all proofs, it is sufficiently demonstrated, since simple motions are but three: to the center, from the center, and about the center, of which the two straight, up and down, are manifestly contrary; and, because one thing has only one contrary, there remains no other motion which may be contrary to the circular. You see the subtle and most conclusive discourse of Aristotle whereby he proves the incorruptibility of heaven.

SALV.: This is nothing more than the same method of Aristotle that I outlined before. As soon as I affirm that the motion which you attribute to the celestial bodies applies also to the Earth, the argument proves nothing. The circular motion which you assign to celestial bodies also suits the Earth, from which, supposing the rest of your discourse to be conclusive will follow one of three things, as I told you a little before and shall repeat. Either the Earth itself is ungenerated and incorruptible like the celestial bodies; or the celestial bodies are like the elementary—generated, alterable, etc.; or this difference of motion has nothing to do with generation and corruption. Aristotle's discourse, and yours also, contains many propositions that might be more closely examined; to do so, it will be necessary to make them as distinct and definite as possible. Excuse me, therefore, Sagredo, if perhaps with some tediousness you hear me often repeat the same things and take up the argument as if in the public circle of disputations. You say that "generation and corruption are only made where there are contraries; contraries being only among simple natural bodies, movable with contrary motions; contrary motions are only those which are made by a straight line between contrary places; and these are only two, from the center and to the center; and such motions belong to no other natural bodies than to earth and fire and the other two elements; therefore, generation and corruption are only among the elements. And as the third simple motion, namely, the circular about the center, has no contrary, therefore that natural body which has such a motion wants a contrary; and having no contrary is ungenerated, and incorruptible, etc., for where there is no contrary, there is no generation or corruption, etc. But such motion applies only with the celestial bodies; therefore, only these are ungenerated and incorruptible, etc."

To begin, I think it easier and quicker to decide whether the Earth (a most vast body, and for its vicinity to us, most tractable) moves with a speedy motion, such as its revolution about its own axis every twenty-four hours would be, than it is to understand and resolve whether generation and corruption arise from contrariety, or else whether there be such things as generation, corruption, and contrariety in Nature. And if you, Simplicio, can tell me what method Nature observes in working when she in a very short time begets an infinite number of flies from a little vapour of the must of wine, and can show me where are the contraries you speak of, what it is that corrupts, and how, I would think even more of you than I do at present; for I confess I cannot comprehend these things. Besides, I would very gladly understand how and why these corruptive contraries are so favorable to daws, and so cruel to doves; so indulgent to stags, and so hasty to horses, that they grant the former many more years of life, that is, of incorruptibility, than weeks to the latter. Peaches and olives are planted in the same soil, exposed to the same heat and cold, to the same wind and rains, and, in a word, to the same contraries, and yet the former decay in a short time, and the latter live many hundred years. Furthermore, I never was thoroughly satisfied about this substantial transmutation (still keeping within pure natural bounds) whereby a matter becomes so transformed that it is necessarily said to be destroyed, so that nothing remains of its first being, and another body quite differing from it should be then produced. If I fancy to myself a body under one aspect, and by-and-by under another very different, I cannot think it impossible but that it may happen by a simple transposition of parts without corrupting or engendering anything anew; for we see such kinds of metamorphosis daily. To return to my purpose, I answer you that, if you go about to persuade me, by way of generability and corruptibility, that the Earth cannot move circularly, you have undertaken a much harder task than I, who with arguments more difficult, indeed, but also more conclusive tend to prove the contrary. . . .

[Galileo's insistence that mathematical demonstrations provide knowledge as certain as that which the Divine Wisdom knows.]

SAGR.: I always accounted as extraordinarily foolish those who would make human comprehension the measure of what Nature has a power or knowledge to effect, whereas on the contrary there is not any least effect in Nature which can be fully understood by the most speculative minds in the world. Their vain presumption of knowing all can take beginning solely from their never having known anything; for if one has but once experienced the perfect knowledge of one thing, and truly tasted what it is to know, he shall perceive that of infinite other conclusions he understands not so much as one.

SALV.: Your discourse is very concluding; for confirmation we have the example of those who know, or have known, something; the wiser they are, the more they know and freely confess that they know little; nay, the wisest man in all Greece, and pronounced as such by the Oracle, openly professed to know that he knew nothing.

SIMP.: It must be confessed therefore that either Socrates or the Oracle itself was a liar, the latter declaring him to be most wise, and he confessing that he knew himself to be most ignorant.

SALV.: Neither one nor the other does follow, for both the assertions may be true. The Oracle adjudges Socrates the wisest among men, whose knowledge is limited; Socrates acknowledges that he knows nothing in relation to absolute wisdom, which is infinite; and, because of the infinite, much is the same part as is little and as is nothing (for to arrive, e.g., to the infinite number, it is all one to accumulate thousands, tens, or zeros) therefore, Socrates well perceived his wisdom to be nothing in comparison with the infinite knowledge that he lacked. But yet, because there is some knowledge found among men, and this not equally shared by all, Socrates might have a greater share thereof than others and therefore verified the answers of the Oracle.

SAGR.: I think I very well understand this particular. Among men, Simplicio, there is the power of influence, but not equally dispensed to all; it is without question that the power of an emperor is far greater than that of a private person, but both the former and the latter are nothing in comparison with the Divine Omnipotence. Amongst men there are some who better understand agriculture than many others; but what has the knowledge of planting a vine in a trench to do with the knowledge that it takes to make it sprout forth, to attract nourishment, select this good part from that other to make of it leaves, another part to make sprouts, another to make grapes, another to make seeds, another to make the husks of them, for such are the works of most wise Nature? This is only one particular operation of the innumerable ones which Nature carries out, and it alone is enough to reveal an infinite wisdom, so that Divine Wisdom may be concluded to be infinitely infinite.

SALV.: Take thereof another example. Do we not say that the skillful revealing of a most lovely statue in a piece of marble has sublimated the wit of Michelangelo Buonarroti far above the vulgar wits of other men? And yet this work is only the imitation of a mere aptitude and disposition of exterior and superficial members of an immovable man; but what is it in comparison of a man made by Nature, composed of so many exterior and interior members, of so many muscles, tendons, nerves, bones, which serve to so many and sundry motions? But what shall we say of the senses, and of the powers of the soul, and, lastly, of the understanding? May we not say, and with reason, that the structure of a statue falls far short of the formation of a living man, yea, more, of a contemptible worm?

SAGR.: And what difference do you think was there between the dove of Archytas and one made by Nature?

SIMP.: Either I am not one of the men who understand, or else there is a manifest contradiction in this your discourse. You account understanding the chief distinction ascribed to man, who is made by Nature, and a little earlier you said, with Socrates, that he had no knowledge at all; therefore you must say that neither did Nature understand how to make an understanding that understands.

SALV.: You argue very cunningly, but, to reply to your objection, we should have recourse to a philosophical distinction and say that the understanding is to be taken two ways, that is, *intensively* or *extensively*. *Extensively*, that is, as to the multitude of intelligibles, which are infinite, the understanding of man is as nothing, though he should understand a thousand propositions; for a thousand in respect of infinity is but as zero. But as for the understanding *intensively*, inasmuch as that term imports perfectly some propositions, I say that human wisdom understands some propositions as perfectly and is as absolutely certain thereof, as Nature herself; and such are the pure mathematical sciences, to wit, Geometry and Arithmetic. In these Divine Wisdom knows infinitely more propositions, because it knows them all; but I believe that the knowledge of those few comprehended by human understanding equals the Divine, as to objective certainty, for it arrives to comprehend the necessity of it, than which there can be no greater certainty.

SIMP.: This seems to me a very bold and rash expression.

SALV.: These are common notions far from all umbrage of temerity, or boldness, and detract not in the least from the majesty of Divine Wisdom, as it in no ways diminishes its omnipotence to say that God cannot make what once happened not to have happened. I believe, Simplicio, that your scruple arises from your having possibly misunderstood my words somewhat; therefore, the better to express myself, I say that as concerns the truth, of which mathematical demonstrations give us the knowledge, it is the same as that which the Divine Wisdom knows. But this I must grant you, that the manner whereby God knows the infinite propositions of which we understand some few is much more excellent than ours, which proceeds by ratiocination and passes from conclusion to conclusion, whereas His is done at one single thought or intuition. For example, we, to attain the knowledge of some property of the circle, which has infinitely many, begin from one of the most simple and, taking that for its definition, do proceed with argumentation to another, and from that to a third, and then to a fourth, and so on. The Divine Wisdom by the simple apprehension of its essence comprehends, without temporal ratiocination, all these infinite properties which are also, in effect, virtually comprised in the definitions of all things; and, to conclude, being infinite, are perhaps but one alone in their nature and in the Divine Mind. Neither is this wholly unknown to human understanding, but only beclouded with deep and dense mists which in part come to be dissipated and clarified when we are made masters of any conclusions firmly demonstrated and made so perfectly ours that we can speedily run through them. For, in sum, what else is that proposition, that the square of the side subtending the right angle in any triangle is equal to the squares of the other two which include it, but that parallelograms built on common bases, and between parallels, are equal amongst themselves? And this, lastly, is it not the same as to say that those two areas are equal when superposed they coincide? Now these inferences, which our intellect apprehends with time and gradual motion, the Divine Wisdom, like light, penetrates in an instant, which is the same as to say has them always all present. I conclude therefore that our understanding, as to both the manner and the multitude of the things comprehended by us, is infinitely surpassed by the Divine Wisdom; but yet I do not so vilify it as to consider it absolutely nothing; rather, when I consider how many and how great mysteries men have understood, discovered, and contrived, I very plainly know and understand the mind of man to be one of the works of God, yea, one of the most excellent.

[His repudiation of traditional experiments purporting to demonstrate the immobility of the earth.]

SALV.: . . . Aristotle says that it is a most convincing argument of the Earth's immobility to see that projectiles thrown or shot upright return perpendicularly by the same line unto the same place from whence they were shot or thrown. And this holds true, although the motion be of a very great height. So that hither may be referred the argument taken from a shot fired directly upwards from a cannon, as also that other used by Aristotle and Ptolemy, of the heavy bodies that, falling from on high, are observed to descend by a direct and perpendicular line to the surface of the Earth. Now, that I may begin to untie these knots, I demand this of Simplicio: in case one should deny to Ptolemy and Aristotle that weights in falling freely from on high descend by a right and perpendicular line, that is, directly to the center, what means would he use to prove it?

SIMP.: The means of the senses, which assure us that that tower or other altitude is upright and perpendicular, and show us that that stone slides along the wall, without inclining a hair's breadth to one side or another, and lights on the ground just under the place from where it was let fall.

SALV.: But if it should happen that the terrestrial globe did move round, and consequently the tower also along with it, and that the stone did then also graze and slide along the side of the tower, what must its motion be then?

SIMP.: In this case we may rather say its motions, for it would have one wherewith to descend from the top to the bottom and should then have another to follow the course of the said tower.

SALV.: So that its motion should be compounded of two; from this it would follow that the stone would no longer describe that simple straight and perpendicular line but one transverse and perhaps not straight.

SIMP.: I can say nothing of its nonrectitude, but this I know very well: that it would of necessity be transverse.

SALV.: You see then that, merely observing the falling stone to glide along the tower, you cannot certainly affirm that it describes a line which is straight and perpendicular unless you first suppose that the Earth stands still.

SIMP.: True, for, if the Earth should move, the stone's motion would be transverse and not perpendicular.

SALV.: Then will you please hold the paralogism of Aristotle and Ptolemy to be evident and manifest, and discovered by yourself, for in it that is supposed for known which is intended to be demonstrated.

SIMP.: How can that be? To me it appears that the syllogism is rightly demonstrated without petition of principle.

SALV.: You shall see how it is; first answer me, does he not lay down the conclusion as unknown?

SIMP.: Unknown: why, otherwise, demonstrating it would be superfluous.

SALV.: But the middle term, ought not that to be known?

SIMP.: It is necessary that it should; for otherwise it would be proving *ignotum per aeque ignotum,* the unknown by the equally unknown.

SALV.: Our conclusion which is to be proved, and which is unknown, is it not the stability of the Earth?

SIMP.: It is.

SALV.: The middle term, which ought to be known, is it not the straight and perpendicular descent of the stone?

SIMP.: It is so.

SALV.: But was it not just now concluded that we can have

no certain knowledge whether it shall be direct and perpendicular unless we first know that the Earth stands still? Therefore in your syllogism the certainty of the middle term is assumed from the uncertainty of the conclusion. You may see, then, what and how great the paralogism is.

SAGR.: I would defend Aristotle in favor of Simplicio. Should the tower move, it would be impossible that the stone should fall gliding along the side of it; and therefore from its falling in that manner the stability of the Earth is inferred.

SIMP.: It is so; for if you would have the stone in descending to graze along the tower, while being carried around by the Earth, you must allow the stone two natural motions, to wit, the straight motion toward the center and the circular motion about the center, which is impossible.

SALV.: Aristotle's defense then consists in the impossibility, or at least in his esteeming it an impossibility, that the stone should move with a motion mixed of right and circular. For, if he did not hold it impossible that the stone could move at once to the center and about the center, he would have understood that it might come to pass that the falling stone might in its descent graze the tower as well when it moved as when it stood still. Consequently, he ought to have perceived that from this grazing nothing could be inferred touching the mobility or immobility of the Earth. But this does not in any way excuse Aristotle; because he ought to have expressed it, if he had had such a notion, it being so material a part of his argument. Also because it cannot be said that such an effect is impossible or that Aristotle did esteem it so. The first cannot be affirmed, for by and by I shall show that it is not only possible but necessary; nor can the second be averred, for Aristotle himself grants that fire moves naturally upwards in a right line, and moves about with the diurnal motion, imparted by the heavens to the whole element of fire and the greater part of the upper air. If therefore he held it possible to mix the straight motion upwards with the circular communicated to the fire and air from the concave of the sphere of the Moon, much less ought he to account impossible the mixture of the straight motion of the stone downwards with the circular which we presuppose natural to the whole terrestrial globe, of which the stone is a part.

SIMP.: I see no such thing; for, if the element of fire revolves round together with the air, it is a very easy, even a necessary thing that a spark of fire which mounts upwards from the Earth, in passing through the moving air, should receive the same motion, being a body so thin, light, and easy to be moved. But that a very heavy stone, or a cannon ball, that descends from on high, and that is at liberty to move whither it will, should suffer itself to be transported either by the air of any other thing is altogether incredible. Besides that, we have the experiment which is so proper to our purpose, of the stone let fall from the round top of the mast of a ship, which, when the ship lies still, falls at the foot of the mast, but, when the ship moves, falls as far distant from that place, as how far the ship in the time of the stone falling had run forward.

SALV.: There is a great disparity between the case of the ship and that of the Earth, if the terrestrial globe be supposed to have a diurnal motion. For it is manifest that, as the motion of the ship is not natural to it, the motion of all those things that are in it is accidental, whence it is no wonder that the stone which was retained in the round top, being left at liberty, descends downwards without any obligation to follow the motion of the ship. But the diurnal conversion is ascribed to the terrestrial globe for its proper and natural motion, and, consequently, it is so to all parts of the said globe; and, being impressed by Nature, is indelible in them. Therefore that stone that is on the top of the tower has an intrinsic inclination to revolve about the center of its whole in twenty-four hours, and it exercises this same natural instinct eternally, be it placed in any state whatsoever. To be assured of the truth of this, you have to do no more than alter an antiquated impression made in your mind, and to say: as I, hitherto, held it to be the property of the terrestrial globe to rest immovable about its center, and never doubted or questioned but that all particles do also naturally remain in the same state of rest; it is reason, in case the terrestrial globe did move round by natural instinct in twenty-four hours, that the intrinsic and natural inclination of all its parts should also be to follow the same revolution. And thus, without running into any inconvenience, one may conclude that, since the motion conferred by the force of oars on the ship, and by it on all the things that are contained within her, is not natural but foreign, it is very reasonable that that stone, being separated from the ship, should bring itself back to its natural disposure and return to exercise its pure simple instinct given it by Nature. To this I add that it is necessary that at least that part of the air which is beneath the greater heights of mountains should be transported and carried round by the roughness of the Earth's surface; or that, as being mixed with many vapors and terrene exhalations, it should naturally follow the diurnal motion, which does not occur in the air about the ship rowed by oars. Therefore your arguing from the ship to the tower has not the force of a conclusion, because the stone which falls from the round top of the mast enters into a medium which is unconcerned in the motion of the ship; but that which departs from the top of the tower finds a medium that has a motion in common with the whole terrestrial globe, so that, rather than being hindered, it is assisted by the motion of the air and may follow the universal course of the Earth.

SIMP.: I cannot conceive that the air can imprint in a very great stone, or in a heavy globe of wood or ball of lead, as suppose of two hundred-weight, the motion wherewith itself is moved, and which it perhaps communicates to feathers, snow, and other very light things. On the contrary, I see that a weight of that nature, being exposed to the most impetuous wind, is not thereby removed an inch from its place; now consider with yourself whether the air will carry it along therewith.

SALV.: There is a great difference between your experiment and our case. You introduce the wind blowing against that stone, supposedly in a state of rest, and we expose the stone to the air which already moves, with the same velocity as the stone, so that the air is not to confer a new motion upon it but only to maintain or, to say better, not to hinder the motion already acquired. You would drive the stone with a strange and preternatural motion, and we desire to conserve it in its natural one. If you would produce a more pertinent experiment, you should say that it is observed, if not with the eye of the forehead, yet with that of the mind, what would happen if an eagle that is carried by the course of the wind should let a stone fall from its talons. Now as the stone went along with the wind when it was let go, and after it started on its fall entered into a medium that moved with equal velocity, I am very confident that it would not be seen to descend in its fall perpendicularly but would move with a transverse motion.

SIMP.: But it should first be known how such an experiment may be made, and then one might judge according to the event. In the meantime the effect of the ship hitherto inclines to favor our opinion.

SALV.: Well you said "hitherto," for perhaps it may change countenance anon. And that I may no longer hold you in suspense, tell me, Simplicio, do you really believe that the experiment of the ship squares so very well with our purpose that it ought to be believed that that which we see happen in it ought also to take place in the terrestrial globe?

SIMP.: As yet I am of that opinion; and, though you have alleged some small disparities, I do not think them of so great a moment that they should make me change my judgment.

SALV.: I rather desire that you would continue in that

belief and hold for certain that the effect of the Earth would exactly answer that of the ship, provided that when it shall appear prejudicial to your cause, you will not be of humour to alter your thoughts. You say that, when the ship stands still, the stone falls at the foot of the mast and that, when she is under sail, it lights far from thence; therefore, by conversion, from the stone's falling at the foot is inferred the ship's standing still, and from its falling far from thence is inferred her moving. And because that which occurs to the ship ought likewise to befall the Earth, therefore from the falling of the stone at the foot of the tower is necessarily inferred the immobility of the terrestrial globe. Is this not your argumentation?

SIMP.: It is; and reduced into such conciseness that it is become most easy to be apprehended.

SALV.: Now tell me; if the stone let fall from the round top when the ship is in swift course, should fall exactly in the same place of the ship in which it falls when the ship is at anchor, what service would these experiments do you, to the end of ascertaining whether the vessel does stand still or move?

SIMP.: Exactly none. As, for example, from the beating of the pulse one cannot know whether a person is asleep or awake, seeing that the pulse beats in the same manner in sleeping as in waking.

SALV.: Very well. Have you ever tried the experiment of the ship?

SIMP.: I have not; but yet I believe that those authors who allege the same have accurately observed it; besides that, the cause of the disparity is so manifestly known that it admits of no questions.

SALV.: It may be that those authors believed in it, without having made trial of it, and you yourself are a good witness to the point. For you, without having examined it, allege it as certain and in good faith rely on their authority; as it is now not only possible but obvious that they also relied on their predecessors, without ever arriving at one that had made the experiment. For, mark you, whosoever shall perform it shall find the event succeeded quite contrary to what has been written of it. That is, he shall see the stone fall at all times in the same place of the ship, whether it stand still or move with any velocity whatsoever. So that, the same holding true in the Earth as in the ship, one cannot, from the stone's falling perpendicularly at the foot of the tower, conclude anything touching the motion or rest of the Earth. . . .

[Examination of the celestial phenomena, which, Galileo says, "make for the Copernican hypothesis, as if it were to prove absolutely victorious; adding by the way certain new observations which yet serve only the astronomical facility, not for natural necessity."]

Simp.: But from whence do you argue that not the Earth but the Sun is in the center of the planetary revolutions?

Salv.: I infer the same from most evident and therefore necessarily conclusive observations, of which the most potent to exclude the Earth from the said center, and to place the Sun therein, are that we see all the planets sometimes nearer and sometimes farther off from the Earth, with so great differences, that, for example, Venus when it is at the farthest is six times more remote from us than when it is nearest, and Mars rises almost eight times as high at one time as at another. See therefore whether Aristotle was not somewhat mistaken in thinking that it was at all times equidistant from us.

Simp.: What in the next place are the tokens that their motions are about the Sun?

Salv.: It is shown in the three superior planets, Mars, Jupiter, and Saturn, in that we find them always nearest to the Earth when they are in opposition to the Sun and farthest off when they are toward the conjunction; and this approximation and recession imports thus much, that Mars near at hand appears sixty times greater than when it is remote. As to Venus, in the next place, and to Mercury, we are certain that they revolve about the Sun in that they never move far from it, and in that we see them sometimes above and sometimes below it, as the mutations of figure in Venus necessarily prove. Touching the Moon, it is certain that it cannot in any way separate itself from the Earth, for the reasons that shall be more distinctly alleged hereafter.

SAGR.: I expect that I shall hear more admirable things that depend upon this annual motion of the Earth than were those dependent upon the diurnal revolution.

SALV.: You are not wrong, for, as to the operation of the diurnal motion upon the celestial bodies, it neither was nor can be other than to make the Universe seem to run precipitately the contrary way; this annual motion, on the other hand, intermixing with the particular motions of all the planets, produces very many anomalies which have disarmed and nonplussed all the greatest scholars in the world. Returning, however, to our first general apprehensions, I reply that the center of the celestial conversions of the five planets, Saturn, Jupiter, Mars, Venus, and Mercury, is the Sun and shall be likewise the center of the motion of the Earth if we do but succeed in our attempt of placing it in heaven. And, as for the Moon, it has a circular motion about the Earth, from which (as I said before) it can by no means alienate itself, but yet it does not cease to go about the Sun together with the Earth in an annual motion.

SIMP.: I do not as yet very well apprehend this structure, but it may be that with a drawing one may better and more easily discourse concerning the same.

SALV.: Good; and, indeed, for your greater satisfaction and astonishment, I want you to draw it yourself and to see that, although you think you do not apprehend it, yet you understand it very perfectly; and only by answering to my interrogations you shall design it exactly. Take therefore a sheet of paper and compasses. Let this white paper be the immense expanse of the Universe, in which you are to distribute and dispose its parts in order, according as reason shall direct you. And, first, since without my instruction you verily believe that the Earth is placed in this Universe, therefore note a point at pleasure, about which you intend it to be placed, and mark it with some characters.

SIMP.: Let this mark *A* be the place of the terrestrial globe.

SALV.: Very well. I know, secondly, that you understand perfectly that the Earth is not within the body of the Sun, nor so much as contiguous to it, but distant some space from the same; therefore, assign to the Sun what other place you like, as remote from the Earth as you please, and mark this in like manner.

SIMP.: Here it is; let the place of the solar body be *O*.

SALV.: These two being constituted, we should think of placing Venus in such a manner that its state and motion may agree with what the experience of the senses shows us and therefore recall to mind that which, either by the past discourses or your own observations, you have learnt to befall that star and afterwards assign to it that state which you think agrees with the same.

SIMP.: Supposing those appearances expressed by you, and which I have likewise read in the little treatise of Conclusions, to be true, namely, that that star never recedes from the Sun beyond a certain interval of 40° or thereabouts, so that it never comes either to opposition with the Sun, or so much as to quadrature, or yet to the sextile aspect; and, more than that, supposing that it appears at one time almost forty times greater than at another, namely, very great, when being retrograde it goes to the vespertine conjunction of the Sun, and very small when, with a motion straight forwards, it goes to the matutine conjunction; and, moreover, it being true that, when it appears big, it shows with a corniculate or horned figure, and, when it appears little, it seems perfectly round—these appearances, I say, being true,

I do not see how one can choose but affirm the said star to revolve in a circle about the Sun, for that the said circle cannot in any wise be said to encompass or to contain the Earth within it, nor to be inferior to the Sun, that is, between it and the Earth, nor yet superior to the Sun. That circle cannot encompass the Earth, because Venus would then sometimes come to opposition with the Sun; it cannot be inferior, for then Venus in both its conjunctions with the Sun would seem horned; nor can it be superior, for then it would always appear round and never cornicular; and therefore I will draw for Venus the circle *CH* about the Sun, without encompassing the Earth.

SALV.: Having placed Venus, it is requisite that you think of Mercury, which, as you know, always keeping about the Sun, does recede less distance from it than Venus; therefore, consider with yourself what place is most convenient to assign it.

SIMP.: It is not to be questioned but that, this planet imitating Venus, the most proper place for it will be a lesser circle within this of Venus, in like manner about the Sun; we may therefore upon these considerations draw its circle, marking it with the characters *BG*.

SALV.: But Mars, where shall we place it?

SIMP.: Mars, because it comes to an opposition with the Sun, its circle must of necessity encompass the Earth; but I see that it must necessarily encompass the Sun also, for coming to conjunction with the Sun, if it did not move over it but were below it, it would appear horned, as Venus and the Moon; but it shows always round, and therefore it is necessary that it should no less include the Sun within its circle than the Earth. And because I remember that you did say that when it is in opposition with the Sun it seems sixty times bigger than when it is in the conjunction, I think that a circle about the center of the Sun that takes in the Earth will very well agree with these phenomena; which

I note and mark *DI*, where Mars in the point *D* is near to the Earth and opposite to the Sun: but, when it is in the point *I*, it is at conjunction with the Sun but very far from the Earth. And because the same appearances are observed in Jupiter and Saturn, although with much lesser difference in Jupiter than in Mars, and with yet less in Saturn than in Jupiter, I understand that we should very aptly save all the phenomena of these two planets, with two circles in like manner drawn about the Sun, and this first for Jupiter, marking it *EL*, and another above that for Saturn marked *FM*.

SALV.: You have behaved yourself bravely hitherto. And because (as you see) the measure of the approach and recession of the three superior planets is given by double the distance between the Earth and the Sun, this makes greater difference in Mars than in Jupiter, the circle *DI* of Mars being lesser than the circle *EL* of Jupiter; and likewise because this *EL* is less than this circle *FM* of Saturn, the said difference is also yet lesser in Saturn than in Jupiter, and that exactly answers the phenomena. It remains now that you assign a place to the Moon.

SIMP.: Following the same method (which seems to me very conclusive), since we see that the Moon comes to conjunction and opposition with the Sun, it is necessary to say that its circle encompasses the Earth, but yet it does not follow that it must environ the Sun, for then at the time of its conjunction it would not seem it could ever eclipse the Sun, as it often does; it is necessary therefore to assign it a circle about the Earth, which should be this, *NP*, so that, being placed in *P*, it will appear from the Earth *A* to be in conjunction with the Sun, and, placed in *N*, it appears oppsite to the Sun, and in that position it may fall under the Earth's shadow and be obscured.

SALV.: Now, Simplicio, what shall we do with the fixed stars? Shall we suppose them scattered through the immense abysses of the Universe, at different distances from one determinate point; or else placed in a surface spherically distended about a center of its own, so that each of them may be equidistant from the said center?

SIMP.: I would rather take a middle way and would assign them a circle described about a determinate center and comprised within two spherical surfaces, to wit, one very high and concave, and the other lower, and convex, betwixt which I would constitute the innumerable multitude of stars, but yet at diverse altitudes, and this might be called the sphere of the Universe, containing within it the circles of the planets already by us described.

SALV.: But now we have all this while, Simplicio, disposed the mundane bodies exactly according to the order of Copernicus, and we have done it with your hand; and moreover to each of them you have assigned peculiar motions of their own, except to the Sun, the Earth, and starry sphere; and to Mercury with Venus you have ascribed the circular motion about the Sun, without encompassing the Earth; about the same Sun you make the three superior planets, Mars, Jupiter, and Saturn, to move, comprehending the Earth within their circles. The Moon in the next place can move in no other manner than about the Earth, without taking in the Sun, and in all these motions you agree also with Copernicus. There remain now three things to be decided between the Sun, the Earth, and fixed stars, namely, *rest*, which seems to belong to the Earth; the *annual motion* under the zodiac, which appears to pertain to the Sun; and the *diurnal motion,* which seems to belong to the starry sphere and to be imparted by it to all the rest of the Universe, the Earth excepted. And it being true that all the circles of the planets, I mean Mercury, Venus, Mars, Jupiter, and Saturn, do move about the Sun as their center, rest seems with so much more reason to belong to the said Sun than to the Earth, inasmuch as in a movable sphere it is more reasonable that the center stand still than any other place remote from the center; to the Earth, therefore, which is constituted in the midst of movable parts of the Universe, I mean between Venus and Mars, one of which makes its revolution in nine months and the other in two years, the motion of a year may very aptly be assigned, leaving rest to the Sun. And, if that be so, it follows of necessary consequence that, likewise, the diurnal motion belongs to the Earth; for if, the Sun standing still, the Earth should not revolve about itself but have only the annual motion about the Sun, our year would be no other than one day and one night, that is, six months of day and six months of night, as has already been said. You may consider withal how aptly the precipitate motion of twenty-four hours is taken away from the Universe, and the fixed stars, that are so many Suns, are made in conformity to our Sun to enjoy a perpetual rest. You see, moreover, what facility one meets with in this rough drawing to render the reason of so great appearances in the celestial bodies.

SAGR.: I very well perceive that facility, but, as you collect, from this simplicity, great probabilities for the truth of that system, others haply could make thence contrary deductions: wondering, not without reason, why being the ancient system of the Pythagoreans, and so well corresponding to the phenomena, it has in the succession of so many thousand years had so few followers and has been refuted even by Aristotle himself, and why later Copernicus himself has had no better fortune.

SALV.: My dear Sagredo, if it has ever been your fate, as it has been mine, many and many a time, to hear what kind of idiocies are enough to make the vulgar contumacious and refractory, I will not say to agreeing but even to listening to these new ideas, I believe that your wonder at the paucity of those who are followers of that opinion would be much diminished. But small regard, in my judgment, ought to be had of such thick souls as think it a most convincing proof to confirm and steadfastly settle them in the belief of the Earth's immobility to see that in the same day they cannot dine at Constantinople and sup in Japan, and that the Earth, as being a most grave body, cannot clamber above the Sun and then slide headlong down again. Of such as these, whose number is infinite, we need not make any reckoning, nor need we to record their fooleries or strive to gain to our side, as partakers in subtle and difficult opinions, individuals in whose definition the kind only is concerned and the difference is wanting. Moreover, what ground do you think you could be able to gain, with all the demonstrations of the world, upon brains so stupid as are not able of themselves to know their utter follies? But my wondering, Sagredo, is very different from yours. You wonder that so few are followers of the Pythagorean opinion; and I am amazed how there could be any yet left till now that do embrace and follow it. Nor can I sufficiently admire the eminence of those men's intelligence who have received and held it to be true, and with the sprightliness of their judgments offered such violence to their own senses that they have been able to prefer that which their reason dictated to them to what sensible appearances represented most manifestly on the contrary. That the reasons against the diurnal vertiginous revolution of the earth, by you already examined, do carry great probability with them, we have already seen; as, also, that the Ptolemaics and Aristotelians with all their sectators did receive them for true is indeed a very great argument of their efficacy; but those experiences which overtly contradict the annual motion have yet so much more of an appearance of convincingness that (I say it again) I cannot find any bounds for my admiration how reason was able in Aristarchus and Copernicus to commit such a rape upon their senses as, in despite thereof, to make herself mistress of their belief.

SAGR.: Are we then to have still more of these strong oppositions against this annual motion?

SALV.: We are, and they are so evident and sensible that, if a sense more sublime and excellent than those common and vulgar did not take part with reason, I much fear that

The Copernican Controversy

I also should have been much more averse to the Copernican system than I have been, since the time that a clearer lamp than ordinary has enlightened me.

SAGR.: Now, therefore, Salviati, let us come to join battle, for every word that is spent on anything else I take to be cast away.

SALV.: I am ready to serve you. You have already seen me draw the form of the Copernican system; now against the truth of it Mars itself, in the first place, makes a hot charge. For in case it were true that its distances from the Earth should so much vary that from the least distance to the greatest there were twice as much difference as from the Earth to the Sun, it would follow that, when it is nearest to us, it should show more than sixty times bigger than it seems when it is farthest from us. Nevertheless, that diversity of apparent magnitude is not to be seen; nay, in its opposition with the Sun, when it is nearest to the Earth, it does not show so much as quadruple and quintuple in bigness to what it is when toward the conjunction it comes to be occulted under the Sun's rays. Another and greater difficulty does Venus exhibit; for if revolving about the Sun, as Copernicus affirms, it were sometimes above, and at other times below it, receding and approaching with respect to us so much as the diameter of the circle described would be, then at such time as it is below the Sun, and nearest to us, its disk would show little less than forty times bigger than when it is above the Sun, near to its other conjunction; yet, nevertheless, the difference is almost imperceptible. Let us add another difficulty: that in case the body of Venus is of itself dark and only shines, as the Moon, by the illumination of the Sun, which seems most reasonable, it would show forked or horned on such time as it is under the Sun, as the Moon does when it is in like manner near the Sun, an accident that is not to be discovered in it. Whereupon Copernicus affirms that, either the Moon is light of itself or else that its substance is of such nature that it can imbue the solar light and transmit the same through all its whole depth, so as to be able to appear to us always shining. And in this manner Copernicus excuses the not changing of figure in Venus; but of its small variation of magnitude, he makes no mention at all and much less of Mars than was needful; I believe as being unable to save so well as he desired a phenomenon so contrary to his hypothesis. Yet, being convinced by so many other occurrences and reasons, he maintained and held the same hypothesis to be true. Besides, to make the planets, together with the Earth, move about the Sun as the center of their conversions, and the Moon alone break that order and have a motion by itself about the Earth; and then to make both it, the Earth, and the whole sphere of the elements, move all together about the Sun in a year, this seems to pervert the order of this system and to render it unlikely and false. These are those difficulties that make me wonder how Aristarchus and Copernicus, who must needs have observed them, not having been able to save them, have yet notwithstanding by other admirable occurrences been induced to confide so much in that which reason dictated to them as to affirm confidently that the structure of the Universe could have no other figure than that which they designed to themselves. There are also several other very serious and curious doubts, not so easy to be resolved by the middle sort of wits, but yet penetrated and declared by Copernicus, which we shall postpone till by and by, after we have answered to other objections that seem to make against this opinion. Now, coming to the answers to those three grand objections aforenamed, I say that the two first not only do not contradict the Copernican system but greatly and absolutely favor it; for both Mars and Venus seem unequal to themselves, according to the proportions assigned; and Venus under the Sun seems horned and goes changing figures in itself exactly like the Moon.

SAGR.: But how came this to be concealed from Copernicus and revealed to you?

SALV.: These things cannot be comprehended, save only by the sense of seeing, which by nature was not granted to man so perfect as to be able to attain to the discovery of such differences; nay, even the very instrument of sight is an impediment to itself. But since it has pleased God in our age to vouchsafe to human ingenuity the admirable invention of perfecting our sight by multiplying it four, six, ten, twenty, thirty, and forty times, a vast number of objects that, either by reason of their distance or for their extreme smallness, were invisible to us have by help of the Telescope been rendered visible.

SAGR.: But Venus and Mars are none of the objects invisible for their distance or smallness, yea, we do discern them with our bare natural sight; why then do we not distinguish the differences of their magnitudes and figures?

SALV.: In this, the impediment of our very eye itself has a great share, as but even now hinted, by which the resplendent and remote objects are not represented to us simple and pure but given to us fringed with strange and adventitious rays, so long and dense, that their naked body appears to us increased ten, twenty, a hundred, yea, a thousand times more than it would appear if that mane or thatch of rays were taken away.

SAGR.: Now I remember that I have read something on this subject, I know not whether in the *Solar Letters* or in the *Saggiatore* of our common Friend, but it would be very good, for recalling it into my memory, as also for the information of Simplicio, who, it may be, never saw those writings, if you would explain to us more distinctly how this business stands.

SIMP.: I must confess that all that which Salviati has said is new to me; for, truth is, I never have had the curiosity to read those books, nor have I hitherto given any great credit to the Telescope newly introduced; rather, treading in the steps of other Peripatetic philosophers, my companions, I have thought those things to be fallacies and delusions of the crystals which others have so much admired for stupendous operations. Therefore, if I have hitherto been in error, I shall be glad to be freed from it, and, allured by these novelties already heard from you, I shall the more attentively hearken to the rest.

SALV.: The confidence that these men have in their own capacity is no less unreasonable than the small esteem they have of the judgment of others; and it is strange indeed that they should think themselves able to judge better of such an instrument without ever having made trial of it than those who have made, and daily do make, a thousand experiments of the same. But, I pray you, let us leave this kind of pertinacious men, whom we cannot so much as tax without doing them too great honor. And, returning to our purpose, I say that resplendent objects, whether it is that their light refracts on the humidity that is upon the pupils or that it reflects on the edge of the eyelids, diffusing its reflex rays upon the said pupils, or whether it is for some other reason, do appear to our eye as if they were environed with new rays, and therefore much bigger than their bodies would represent themselves to us were they divested of those irradiations. And this enlargement is made with a greater and greater proportion, by how much those lucid objects are lesser and lesser; in the same manner, for all the world, as if we should suppose that the augmentation of this aura of rays were, for example, 4 inches, which addition being made about a circle that has 4 inches diameter would increase its appearance to nine times its former bigness; but . . .

SIMP.: I believe you meant three times; for adding 4 inches to this side and 4 inches to that side of the diameter of a circle, which is likewise 4 inches, its quantity is thereby tripled and not made nine times bigger.

SALV.: A little more geometry, please, Simplicio. True it is that the diameter is tripled, but the area, which is that of which we speak, increases nine times; for you must know, Simplicio, that the areas of circles are to one another as the squares of their diameters; and a circle that has 4 inches diameter is to another that has 12 as the square of 4 to the square of 12; that is, as 16 is to 144. Therefore, it shall be increased nine times and not three; this, by way of caution to Simplicio. And, proceeding forwards, if we should add said irradiation of 4 inches to a circle that has but 2 inches of diameter only, the diameter of the irradiation or wreath would be 10 inches, and the area of the circle would be to the area of the naked body as 100 to 4, for those are the squares of 10 and of 2. The enlargement would therefore be twenty-five times as much; and, lastly, if the 4 inches of hair or fringe were added to a small circle of an inch in diameter, the same would be increased eighty-one times; and so continually the augmentations are made with a proportion greater and greater, according as the real objects that increase are lesser and lesser.

SAGR.: The doubt which puzzled Simplicio never troubled me, but indeed there are certain other things of which I desire a more distinct understanding; and, in particular, I would know upon what ground you affirm that the said enlargement is always equal in all visible objects.

SALV.: I have already explained it in part when I said that only shining objects increased thus and not the obscure; now I add what remains: that of the resplendent objects, those that are of a more bright light, make the reflection greater and more resplendent upon our pupil, whereupon they seem to augment much more than the less shining; and that I may no more expand upon this particular, let us come to that which the true teacher shows us. Let us this evening, when the air is very dark, observe the star of Jupiter; we shall see it very glittering and very great. Let us afterwards look through a tube, or else through a small chink that, clutching the hand close and accosting it to the eye, we leave between the palm of the hand and the fingers, or else by a hole made with a small needle in a paper; and we shall see the star divested of its beams but so small that we shall judge it less, even than a sixtieth part of its great glittering light seen with the eye at liberty. We may afterwards behold the Dog Star, beautiful and bigger than any of the other fixed stars, which seems to the bare eye not much smaller than Jupiter; but taking from it the irradiation, its disk will show so small that it will not be thought the twentieth part of that of Jupiter, nay, he who has not very good eyes will very hardly discern it; from whence it may be rationally inferred that the said star, as having a much more lively light than Jupiter, makes its irradiation greater than Jupiter does its. As to the irradiation of the Sun and Moon, it is as nothing, because of their magnitude, which occupies of itself alone so great a space in our eye that it leaves no place for the adventitious rays; so that their faces seem close clipped and sharply bounded. We may assure ourselves of the same truth by another experiment which I have often made trial of; we may assure ourselves, I say, that bodies shining with most lively light do irradiate, or beam forth rays, more by far than those that are of a more languishing light. I have many times seen Jupiter and Venus together 20° or 30° distant from the Sun, and, the air being very dark, Venus appeared eight or ten times bigger than Jupiter, being both beheld by the eye at liberty but, being beheld afterwards with the Telescope, the disk of Jupiter showed itself to be four or more times greater than that of Venus, but the vivacity of the splendor of Venus was incomparably bigger than the languishing light of Jupiter, which was only because of Jupiter's being far from the Sun and from us, while Venus is near to us and to the Sun. These things premised, it will not be difficult to comprehend how Mars, when it is in opposition to the Sun and therefore nearer to the Earth by seven times and more than it is toward the conjunction, comes to appear scarce four or five times bigger in that state than in this, where it should appear more than fifty times as much. Only irradiation is the cause; for, if we divest it of the adventitious rays, we shall find it exactly augmented with the due proportion. But, for taking away the light fringe entirely, the Telescope is the best and only means, which by enlarging its disk nine hundred or a thousand times, makes it to be seen naked and terminate, as that of the Moon, and different from itself in the two positions according to its due proportion to an hair. Again, as to Venus, that in its vespertine conjunction, when it is below the Sun, ought to show almost forty times bigger than in the other matutine conjunction, and yet does not appear as much as doubled; it happens, besides the effect of the irradiation, that it is horned; and its crescents, besides being sharp, do receive the Sun's light obliquely and therefore emit but a faint splendor; so that, as being little and weak, its irradiation becomes less ample and vivacious than when it appears to us with its hemisphere all shining. But now the Telescope manifestly shows those horns to be as terminate and distinct as those of the Moon and appear, as it were, as parts of a very large circle, near forty times greater, to judge from them, than the disk of the planet when it is superior to the Sun in its ultimate morning apparition.

SAGR.: Oh, Nicholas Copernicus, how great would have been thy joy to have seen this part of thy system confirmed with so manifest experiences!

SALV.: 'Tis true. But how much less the fame of his greatness among the understanding? When we see, as I said before, that he did constantly continue to affirm (being persuaded thereto by reason) that which sense experience seemed to contradict; for I cannot cease to wonder that he should have constantly persisted in saying that Venus revolves about the Sun, and is more than six times farther from us at one time than at another, while it seems to be always of an equal bigness; although it ought to show forty times bigger when nearest to us than when farthest off.

SAGR.: But in Jupiter, Saturn, and Mercury, I believe that the differences of their apparent magnitudes should seem punctually to answer to their different distances.

SALV.: In the two superior ones, I have made precise observations yearly for this twenty-two years past. In Mercury there can be no observation of moment made because it does not allow itself to be seen, save only in its greatest elongations from the Sun, in which its distances from the Earth are insensibly unequal, and those differences consequently not to be observed; so also its mutations of figures, which must absolutely happen in it as in Venus. If we did see it, it should of necessity appear in form of a semicircle, as Venus does in its greatest digressions; but its disk is so very small, and its splendor so very great, by reason of its vicinity to the Sun, that the virtue of the Telescope does not suffice to clip its tresses or adventitious rays, so as to make them appear shaved round about.

It remains now for us to remove that which seemed a great inconvenience in the motion of the Earth, namely, that, while all the planets move about the Sun, it alone should move round the Sun not solitary, as the rest, but in company with the Moon and the whole sphere of the elements; and that the Moon withal should move every month about the Earth. Here it is necessary once again to exclaim and extol the admirable perspicacity of Copernicus and withal to lament his misfortune in that he is not now alive in our days, when, for removing the seeming absurdity of the Earth and Moon's motion in consort, we see Jupiter, as if it were another Earth, not in consort with the Moon but accompanied by four moons, revolve about the Sun in twelve years together, with whatever things the orbs of the four Medicean Stars can contain within them.

SAGR.: Why do you call those four Jovial planets "moons?"

SALV.: Such they would seem to be to one who standing in Jupiter should behold them; for they are of themselves dark and receive their light from the Sun, which is manifest from their being eclipsed when they enter into the cone of Jupiter's shadow. As only their hemispheres that look toward the Sun are illuminated, to us, who are without their orbs and nearer to the Sun, they seem always lighted, but to one who should be in Jupiter they would show all illuminated at such time as they were in the upper parts of their circles. In the inferior position, that is, between Jupiter and the Sun, they would from Jupiter be observed to be horned; and in a word they would, to the observer standing in Jupiter, make the selfsame changes of figure that the Moon does make to us upon the Earth. You see now how these three things, which at first seemed dissonant, do admirably agree with the Copernican system. Here also, by the way, you may see, Simplicio, with what probability one may conclude that the Sun and not the Earth is in the center of the planetary conversions. And since the Earth is now placed among the bodies that undoubtedly move about the Sun, to wit, above Mercury and Venus and below Saturn, Jupiter, and Mars, shall it not be in like manner probable, and perhaps necessary, to grant that it also moves round?

SIMP.: These accidents are so notable and conspicuous that it is not possible that Ptolemy and other sectators of his should not have had knowledge of them, and, having so, it is likewise necessary that they should have found a way to render reasons of such and so evident appearances that were sufficient, and also congruous and probable, seeing that they have for so long a time been received by such numbers of learned men.

SALV.: You argue very well, but you know that the principal aim of astronomers is to render only reason for the appearances in the celestial bodies and to accommodate such structures and compositions of circles to them that the motions following those calculations answer to the said appearances, little scrupling to admit of some incongruities that indeed upon other accounts they would much stick at.

Copernicus himself writes that he had in his first studies

restored the science of astronomy upon the very suppositions of Ptolemy and in such manner corrected the motions of the planets that the computations did very exactly agree with the phenomena, and the phenomena with the computations, when he took the planets severally one by one. But he adds that, in going about to put together all the structures of the particular orbits, there resulted thence a monster and Chimera, composed of members most disproportionate to one another and altogether incompatible. So that, although it satisfied an merely arithmetical astronomer, yet it did not afford satisfaction or content to the philosophical astronomer, that is, the natural philosopher. And as he very well understood that, if one can save the celestial appearances with false assumptions in Nature, it might with much more ease be done by true suppositions, he set himself diligently to search whether any among the ancient men of fame had ascribed to the world any other structure than that commonly received by Ptolemy; and finding that some Pythagoreans had in particular assigned the diurnal conversion to the Earth, and others the annual motion also, he began to compare the appearances and particularities of the planets' motions with these two new suppositions, all which things jumped exactly with his purpose. Finding the whole to correspond with admirable facility to its parts, he embraced this new system and in it took up his rest.

SIMP.: But what great inconsistencies are there in the Ptolemaic system for which there are not greater to be found in this of Copernicus?

SALV.: In the Ptolemaic hypothesis there are the diseases, and in the Copernican their cure. First, will not all the sects of philosophers account it a great inconvenience that a body naturally movable in circumgyration should move irregularly upon its own center and regularly upon another point? And yet there are such deformed motions as these in the Ptolemean hypothesis, but in the Copernican all move evenly about their own centers. In the Ptolemaic it is necessary to assign to the celestial bodies contrary motions and to make them all move from east to west and at the same time from west to east; but in the Copernican all the celestial revolutions are only one way, from west to east. But what shall we say of the apparent motion of the planets, so irregular, that they not only go one while swift and another while slow, but sometimes wholly cease to move; and then after a long time return back again? To save which appearances Ptolemy introduces very great epicycles, accommodating them one by one to each planet, with various rules of incongruous motions, which are all with one single motion of the Earth taken away. And would you not, Simplicio, call it a great absurdity, if in the Ptolemaic hypothesis, in which the particular planets have their peculiar spheres assigned to them one above another, one finds himself frequently forced to say that Mars, placed above the sphere of the Sun, does so descend that, breaking that sphere, it goes under it and approaches nearer to the Earth than does the body of the Sun and by and by again immeasurably ascends above the same? And yet this and other exorbitancies are remedied by the sole and single motion of the Earth. . . .

SALV.: I could wish, Simplicio, that, suspending for a time the affection that you bear to the followers of your opinion, you would sincerely tell me whether you think that they do in their minds comprehend that magnitude which they reject afterwards as impossible for its immensity to be ascribed to the Universe. For I, as to my own part, think that they do not. But I believe that, as, in the apprehension of numbers, when once a man begins to pass those millions of millions, the imagination is confounded and can no longer form a concept of them, so it happens also in trying to comprehend immense magnitudes and distances; so that there intervenes to the comprehension an effect like to that which befalls the sense. For when in a serene night I look toward the stars, I judge, according to sense, that their distance is but a few miles and that the fixed stars are not a jot more remote than Jupiter or Saturn or than the Moon. But, without more ado, consider the controversies that have passed between the astronomers and the Peripatetic philosophers, upon occasion of the new stars of Cassiopeia and of Sagittarius, the astronomers placing them amongst the fixed stars, and the philosophers believing them to be below the Moon, so unable is our sense to distinguish great distances from the greatest, though these be in reality many thousand times greater than those. In a word, I ask of you, oh, foolish men, does your imagination comprehend that vast magnitude of the Universe, which you afterwards judge to be too immense? If you comprehend it, will you hold that your apprehension extends itself further than the Divine Power? Will you say that you can imagine greater things than those which God can bring to pass? But if you apprehend it not, why will you pass verdict upon things beyond your comprehension?

SIMP.: All this is very well, nor can it be denied but that Heaven may in greatness surpass our imagination, as also that God might have created it thousands of times vaster than now it is; but we ought not to grant anything to have been made in vain and to be idle in the Universe. Now, since we see this admirable order of the planets, disposed about the Earth in distances proportionate for producing their effects for our advantage, to what purpose is to interpose afterwards between the highest sphere of Saturn and the starry sphere a vast vacancy, without any star, that is superfluous and to no purpose? To what end? For whose profit and advantage?

SALV.: I think we arrogate too much to ourselves Simplicio, when we take it for granted that only the care of us is the adequate reason and limit, beyond which Divine Wisdom and Power does or disposes nothing. I will not consent that we should so much shorten its hand, but desire that we may content ourselves with an assurance that God and Nature are so employed in the governing of human affairs that they could not apply themselves more thereto if they truly had no other care than only that of mankind. And this, I think, I am able to make out by a most pertinent and most noble example, taken from the operation of the Sun's light, which, while it attracts these vapors, or heats that plant, attracts and heats them as if it had no more to do; yea, in ripening that bunch of grapes, nay, that one single grape, it does apply itself so that it could not be more intense, if the sum of all its business had been the maturation of that one grape. Now if this grape receives all that it is possible for it to receive from the Sun, not suffering the least injury by the Sun's production of a thousand other effects at the same time, well might we accuse that grape of envy or folly if it should think or wish that the Sun would appropriate all of its rays to its advantage. I am confident that nothing is omitted by the Divine Providence of what concerns the government of human affairs; but that there may not be other things in the Universe that depend upon the same infinite wisdom, I cannot, of myself, by what my reason holds forth to me, bring myself to believe. Surely, I should not forbear to believe any reasons to the contrary laid before me by some higher intellect. But, as I stand, if one should tell me that an immense space interposed between the orbs of the plants and the starry sphere, deprived of stars and idle, would be vain and useless, as likewise that so great an immensity for receipt of the fixed stars as exceeds our utmost comprehension would be superfluous, I would reply that it is rashness to go about to make our shallow reason judge of the works of God, and to call vain and superfluous whatever thing in the Universe is not of use to us.

SAGR.: Say rather, and I believe you would say better, that *we know not what may be its use to us;* and I hold it one of the greatest vanities, yea, follies, that can be in the world, to say, "Because I know not of what use Jupiter or Saturn are to me, therefore these planets are superfluous, yet, more, there are no such things in Nature"; whenas, oh, most foolish men, I do not know so much as to what purpose the arteries, the gristles, the spleen, the gall serve; nay, I should not know that I have a gall, spleen, or kidneys, if in many dissected corpses they were not shown to me; and only then shall I be able to know what the spleen works in me when it comes to be taken from me. To be able to know what this or that celestial body works in me (since you will have it that all their influences direct themselves to us), it would be requisite to remove that body for some time; and then, whatever effect I should find wanting in me, I would say that it depended on that star. Moreover, who will presume to say that the space which they call too vast and useless between Saturn and the fixed stars is void of other mundane bodies? Must it be so, because we do not see them? Then the four Medicean Planets and the companions of Saturn came first into heaven, when we began to see them, and not before? And by this rule the innumerable other fixed stars had no existence before men did look on them? Those cloudy constellations called Nebulae were at first only white flakes, but afterwards with the Telescope we made them to become constellations of many lucid and bright stars. Oh, presumptuous, rather, oh, rash ignorance of man!

SALV.: It is to no purpose, Sagredo, to sally out any more into these unprofitable excursions. Let us pursue our intended design of examining the validity of the reasons alleged on either side, without determining anything, leaving the judgment, when we have done, to such as are more knowing. Returning, therefore, to our natural and human disquisitions, I say that great, little, immense, small, etc., are not absolute but relative terms, so that the selfsame thing, compared with diverse others, may sometimes be called immense and at other times imperceptible, not to say, small. This being so, I demand in relation to what the starry sphere of Copernicus may be called overvast. In my judgment it cannot be compared, or said to be such, unless it be in relation to some other thing of the same kind. Now let us take the very least of the same kind, which shall be the lunar orbit; and, if the starry orb may be so censured to be too big in respect to that of the Moon, every other magnitude that with like or greater proportion exceeds another of the same kind ought to be adjudged too vast and for the same reason to be denied existence in the world; and thus an elephant and a whale shall without more ado be condemned as chimeras and poetical fictions, the one as being too vast in relation to an ant, which is a terrestrial animal, and the other too immeasurable in respect to the gudgeon, which is a fish; yet all of them certainly are seen to be *in rerum natura.* Without any dispute, the elephant and whale exceed the ant and gudgeon in a much greater proportion than the starry sphere does that of the Moon, even if we should fancy the said sphere to be as big as the Copernican system makes it. Moreover, how hugely big is the sphere of Jupiter, or that of Saturn, designed for a receptacle for but one single star and that very small in comparison with one of the fixed? Certainly if we should assign to every one of the fixed stars for its receptacle so great a part of the world's space, it would be necessary to make the orb wherein such innumerable multitudes of them reside very many thousands of times bigger than that which serves the purpose of Copernicus. Besides, do you not call a fixed star very small, I mean even one of the most apparent, and not one of those which shun our sight? And do we not call them so in respect of the vast space circumfused? Now if the whole starry sphere were one entire lucid body, who is there that does not know that in an infinite space there might be assigned a distance so great that the said lucid sphere might from thence show as little, yet, less than a fixed star now appears beheld from the Earth? From thence, therefore, we should *then* judge that selfsame thing to be little which *now* from here we esteem to be immeasurably great.

SAGR.: Great in my judgment is the ineptitude of those

who would have had God to have made the Universe more proportional to the narrow capacities of their reason than to his immense, rather, infinite, power. . . .

[After attempting to demonstrate how the heretofore inexplicable "problem of the tides might receive some light, admitting the Earth's motion," Galileo concludes with a "concession" to the advice of Cardinal Barberini.]

SALV.: Now, since it is time to put an end to our discourses, it remains that I entreat you that, if, going over again at leisure the things that have been alleged, you meet with any doubts or scruples not well resolved, you will excuse my oversight, as well for the novelty of the notion as for the weakness of my intellect as also for the greatness of the subject, as also, finally, because I do not ask nor have pretended to that assent from others which I myself do not give to this theory, which I could very easily grant to be a vain chimera and a most huge paradox. And you, Sagredo, although in the discourses past you have many times, with great applause, declared that you were pleased with some of my conjectures, yet I believe that that was in part more occasioned by the novelty than by the certainty of them, but much more by your courtesy, which did think and desire, by its assent, to procure me that content which we naturally use to take in the approbation and applause of our own matters. And, as your civility has obliged me to you, so am I also pleased with the candor of Simplicio. Nay, his constancy in maintaining the doctrine of his master, with so much strength and undauntedness, has made me much to love him. And, as I am to give you thanks, Sagredo, for your courteous affection, so of Simplicio I ask pardon, if I have sometimes moved him with my too bold and resolute speaking, and let him be assured that I have not done it out of any inducement of sinister affection but only to give him occasion to set before us more lofty thoughts that might make me the more knowing.

SIMP.: There is no reason why you should make all these excuses that are needless, and especially to me, who, being accustomed to be at conferences and public disputes, have a hundred times seen the disputants not only grow hot and angry at one another but break forth into injurious words and sometimes come very near to blows. As for the past discourses, and particularly this last, of the reason of the ebbing and flowing of the sea, I do not, to speak the truth, very well comprehend it. But by that slight idea, whatever it be, that I have formed thereof to myself, I confess that your hypothesis seems to me far more ingenious than any of all those that I ever heard besides; still, I esteem it neither true nor conclusive, but, keeping always before the eyes of my mind a solid doctrine that I once received from a most learned and eminent person, and to which there can be no answer, I know that both of you, being asked whether God, by his infinite power and wisdom, might confer upon the element of water the reciprocal motion in any other way than by making the containing vessel to move, I know, I say, that you will answer that he could, and also knew how to bring it about in many ways, and some of them above the reach of our intellect. Upon which I forthwith conclude that, this being granted, it would be an extravagant boldness for anyone to go about to limit and confine the Divine power and wisdom to some one particular conjecture of his own.

SALV.: An admirable and truly angelical doctrine, which is answered with perfect agreement by that other one, in like manner divine, which gives us leave to dispute touching the constitution of the Universe, but adds, withal (perhaps to the end that the exercise of the minds of men might not cease or become remiss), that we are not to find out the works made by His hands. Let, therefore, the disquisition permitted and ordained us by God assist us in the knowing, and so much more admiring, His greatness by how much less we find ourselves capable of penetrating the profound abysses of His infinite wisdom.

SAGR.: And this may serve for a final close of our four days' disputations, after which, if it seem good to Salviati to take some time to rest himself, our curiosity must, of necessity, grant it to him, yet upon condition that, when it is less inconvenient for him, he will return and satisfy my desire concerning the problems that remain to be discussed and that I have set down to be propounded at one or two other conferences, according to our agreement. And, above all, I shall very impatiently wait to hear the elements of the new science of our Academic about the natural and violent local motions. In the meantime, we may, according to our custom, spend an hour in taking the air in the gondola that waits for us.

The Recantation of Galileo*

I, Galileo, son of the late Vincenso Galilei, Florentine, aged seventy years, arraigned personally before this tribunal and kneeling before you, Most Eminent and Reverend Lord Cardinals, Inquisitors-General against heretical pravity throughout the entire Christian commonwealth, having before my eyes and touching with my hands the Holy Gospels, swear that I have always believed, do believe, and by God's help will in the future believe all that is held, preached, and taught by the Holy Catholic and Apostolic Church. But, whereas—after an injunction had been judicially intimated to me by this Holy Office to the effect that I must altogether abandon the false opinion that the Sun is the center of the world and immovable and that the Earth is not the center of the world and moves and that I must not hold, defend, or teach in any way whatsoever, verbally or in writing, the said false doctrine, and after it had been notified to me that the said doctrine was contrary to Holy Scripture—I wrote and printed a book in which I discuss this new doctrine already condemned and adduce arguments of great cogency in its favor without presenting any solution of these, I have been pronounced by the Holy Office to be vehemently suspected of heresy, that is to say, of having held and believed that the Sun is the center of the world and immovable and that the Earth is not the center and moves:

Therefore, desiring to remove from the minds of your Eminences, and of all faithful Christians, this vehement suspicion justly conceived against me, with sincere heart and

* Giorgio de Santillana, *The Crime of Galileo*. Copyright © 1955 by the University of Chicago Press; reprinted by permission of the University of Chicago Press.

unfeigned faith I abjure, curse, and detest the aforesaid errors and heresies and generally every other error, heresy, and sect whatsoever contrary to the Holy Church, and I swear that in future I will never again say or assert, verbally or in writing, anything that might furnish occasion for a similar suspicion regarding me; but, should I know any heretic or person suspected of heresy, I will denounce him to this Holy Office or to the Inquisitor or Ordinary of the place where I may be. Further, I swear and promise to fulfill and observe in their integrity all penances that have been, or that shall be, imposed upon me by this Holy Office. And, in the event of my contravening (which God forbid!) any of these my promises and oaths, I submit myself to all the pains and penalties imposed and promulgated in the sacred canons and other constitutions, general and particular, against such delinquents. So help me God and these His Holy Gospels, which I touch with my hands.

[Having recited, he signed the attestation as follows]:

I, the said Galileo Galilei, have abjured, sworn, promised, and bound myself as above; and in witness of the truth thereof I have with my own hand subscribed the present document of my abjuration and recited it word for word at Rome, in the convent of the Minerva, this twenty-second day of June, 1633.

I, Galileo Galilei, have abjured as above with my own hand.

CHAPTER *3*

Two New Sciences

All of the selections in this chapter are taken from Galileo's last important work, *Dialogues Concerning Two New Sciences* (1638), which he himself considered to be "superior to everything else of mine hitherto published," containing "results which I consider the most important of all my studies."

As Galileo points out in the third part of this work [see page 145] his new sciences deal with subjects which had been of interest to men from the most ancient times but which, until his own day, no one had studied with precision of observation and accuracy of mathematical formulation. What he actually accomplished in this sphere was, he knew, though considerable in itself, less important than the method by which he accomplished it. In his own estimate his great achievement was that he had "opened up to this vast and most excellent science, of which my work is merely the beginning, ways and means by which other minds more acute than mine will explore its remote corners."

Here we encounter in its perfection the method whereby Galileo enabled his pupils to proceed, by the easiest possible steps, from ordinary observation of commonplace occurrences to precise scientific knowledge.

from the

Dialogues Concerning Two New Sciences*

PUBLISHER'S STATEMENT "TO THE READER" OF THE ORIGINAL EDITION

. . . the divine and natural gifts of this man [Galileo] are shown to best advantage in the present work where he is seen to have discovered, though not without many labors and long vigils, two entirely new sciences and to have demonstrated them in a rigid, that is, geometric, manner: and what is even more remarkable in this work is the fact that one of the two sciences deals with a subject of never-ending interest, perhaps the most important in nature, one which has engaged the minds of all the great philosophers and one concerning which an extraordinary number of books have been written. I refer to motion, a phenomenon exhibiting very many wonderful properties, none of which has hitherto been discovered or demonstrated by any one. The other science which he has also developed from its very foundation deals with the resistance which solid bodies offer to fracture by external forces, a subject of great utility, especially in the sciences and mechanical arts, and one also abounding in properties and theorems not hitherto observed.

In this volume one finds the first treatment of these two sciences, full of propositions to which, as time goes on, able thinkers will add many more; also by means of a large number of clear demonstrations the author points the way to many other theorems as will be readily seen and understood by all intelligent readers.

---

*Galileo, *Dialogues Concerning Two New Sciences,* translated by Henry Crew and Alfonso de Salvio. Reprinted by permission of the Northwestern University Press.

First New Science, Treating the Resistance Which Solid Bodies Offer to Fracture

THE ARSENAL OF THE VENETIANS. RESISTANCE IN THE LITTLE AND THE LARGE

SALVIATI: The constant activity which you Venetians display in your famous arsenal suggests to the studious mind a large field for investigation, especially that part of the work which involves mechanics; for in this department all types of instruments and machines are constantly being constructed by many artisans, among whom there must be some who, partly by inherited experience and partly by their own observations, have become highly expert and clever in explanation.

SAGREDO: You are quite right. Indeed, I myself, being curious by nature, frequently visit this place for the mere pleasure of observing the work of those who, on account of their superiority over other artisans, we call "first rank men." Conference with them has often helped me in the investigation of certain effects including not only those which are striking, but also those which are recondite and almost incredible. At times also I have been put to confusion and driven to despair of ever explaining something for which I could not account, but which my senses told me to be true. And notwithstanding the fact that what the old man told us a little while ago is proverbial and commonly accepted, yet it seemed to me altogether false, like many another saying which is current among the ignorant; for I think they introduce these expressions in order to give the appearance of knowing something about matters which they do not understand.

SALV.: You refer, perhaps, to that last remark of his when we asked the reason why they employed stacks, scaffolding and bracing of larger dimensions for launching a big vessel than they do for a small one; and he answered that they did this in order to avoid the danger of the ship parting under its own heavy weight, a danger to which small boats are not subject?

SAGR.: Yes, that is what I mean; and I refer especially to his last assertion which I have always regarded as a false, though current, opinion; namely, that in speaking of these and other similar machines one cannot argue from the small to the large, because many devices which succeed on a small scale do not work on a large scale. Now, since mechanics has its foundation in geometry, where mere size cuts no figure, I do not see that the properties of circles, triangles, cylinders, cones and other solid figures will change with their size. If, therefore, a large machine be constructed in such a way that its parts bear to one another the same ratio as in a smaller one, and if the smaller is sufficiently strong for the purpose for which it was designed, I do not see why the larger also should not be able to withstand any severe and destructive tests to which it may be subjected.

SALV.: Sagredo, you would do well to change the opinion which you, and perhaps also many other students of mechanics, have entertained concerning the ability of machines and structures to resist external disturbances, thinking that when they are built of the same material and maintain the same ratio between parts, they are able equally, or rather proportionally, to resist or yield to such external disturbances and blows. For we can demonstrate by geometry that the large machine is not proportionately stronger than the small. Finally, we may say that, for every machine and structure, whether artificial or natural, there is set a necessary limit beyond which neither art nor nature can pass; it is here understood, of course, that the material is the same and the proportion preserved.

SAGR.: My brain already reels. My mind, like a cloud momentarily illuminated by a lightning-flash, is for an instant filled with an unusual light, which now beckons to me and which now suddenly mingles and obscures strange, crude ideas. From what you have said it appears to me impossible to build two similar structures of the same material, but of different sizes and have them proportionately strong; and if this were so, it would not be possible to find two single poles made of the same wood which shall be alike in strength and resistance but unlike in size.

SALV.: So it is, Sagredo. And to make sure that we understand each other, I say that if we take a wooden rod of a certain length and size, fitted, say, into a wall at right angles, i.e., parallel to the horizon, it may be reduced to such a length that it will just support itself; so that if a hair's breadth be added to its length it will break under its own weight and will be the only rod of the kind in the world. Thus if, for instance, its length be a hundred times its breadth, you will not be able to find another rod whose length is also a hundred times its breadth and which, like the former, is just able to sustain its own weight and no more: all the larger ones will break while all the shorter ones will be strong enough to support something more than their own weight. And this which I have said about the ability to support itself must be understood to apply also to other tests; so that if a piece of scantling will carry the weight of ten similar to itself, a beam having the same proportions will not be able to support ten similar beams.

Please observe, gentlemen, how facts which at first seem improbable will, even on scant explanation, drop the cloak which has hidden them and stand forth in naked and simple beauty. Who does not know that a horse falling from a height of three or four cubits will break his bones, while a dog falling from the same height or a cat from a height of eight or ten cubits will suffer no injury? Equally harmless would be the fall of a grasshopper from a tower or the fall of an ant from the distance of the moon. Do not children fall with impunity from heights which would cost their elders a broken leg or perhaps a fractured skull? And just as smaller animals are proportionately stronger and more robust than the larger, so also smaller plants are able to stand up better than larger. I am certain you both know that an oak two hundred cubits high would not be able to sustain its own branches if they were distributed as in a tree of ordinary size; and that nature cannot produce a horse as large as twenty ordinary horses or a giant ten times taller than an ordinary man unless by miracles or by greatly altering the proportions of his limbs and especially of his bones, which would have to be considerably enlarged over the ordinary. Likewise the current belief that, in the case of artificial machines the very large and the small are equally feasible and lasting is a manifest error. Thus, for example, a small obelisk or column or other solid figure can certainly be laid down or set up without danger of breaking, while the very large ones will go to pieces under the slightest provocation, and that purely on account of their own weight. And here I must relate a circumstance which is worthy of your attention as indeed are all events which happen contrary to expectation, especially when a precautionary measure turns out to be a cause of disaster. A large marble column was laid out so that its two ends rested each upon a piece of beam; a little later it occurred to a mechanic that, in order to be doubly sure of its not breaking in the middle by its own weight, it would be wise to lay a third support midway; this seemed to all an excellent idea; but the sequel showed that it was quite the opposite, for not many months passed before the column was found cracked and broken exactly above the new middle support.

SIMPLICIO: A very remarkable and thoroughly unexpected accident, especially if caused by placing that new support in the middle.

SALV.: Surely this is the explanation, and the moment the cause is known our surprise vanishes; for when the two pieces of the column were placed on level ground it was observed that one of the end beams had, after a long while, become decayed and sunken, but that the middle one remained hard and strong, thus causing one half of the column to project in the air without any support. Under these circumstances the body therefore behaved differently from what it would have done if supported only upon the first beams; because no matter how much they might have sunken the column would have gone with them. This is an accident which could not possibly have happened to a small column, even though made of the same stone and having a length corresponding to its thickness, i.e., preserving the ratio between thickness and length found in the large pillar.

SAGR.: I am quite convinced of the facts of the case, but I do not understand why the strength and resistance are not multiplied in the same proportion as the material; and I am the more puzzled because, on the contrary, I have noticed in other cases that the strength and resistance against breaking increase in a larger ratio than the amount of material. Thus, for instance, if two nails be driven into a wall, the one which is twice as big as the other will support not only twice as much weight as the other, but three or four times as much.

SALV.: Indeed you will not be far wrong if you say eight times as much; nor does this phenomenon contradict the other even though in appearance they seem so different.

SAGR.: Will you not then, Salviati, remove these difficulties and clear away these obscurities if possible: for I imagine that this problem of resistance opens up a field of beautiful and useful ideas; and if you are pleased to make this the subject of today's discourse you will place Simplicio and me under many obligations.

SALV.: I am at your service if only I can call to mind what I learned from our Academician [in this dialogue "Academician" refers to Galileo himself] who had thought much upon this subject and according to his custom had demonstrated everything by geometrical methods so that one might fairly call this a new science. For, although some of his conclusions had been reached by others, first of all by Aristotle, these are not the most beautiful and, what is more important, they had not been proven in a rigid manner from fundamental principles. Now, since I wish to convince you by demonstrative reasoning rather than to persuade you by mere probabilities, I shall suppose that you are familiar with present-day mechanics so far as it is needed in our discussion. . . .

EFFECTS OF MOTION IN PHENOMENA OF LIGHT AND HEAT. THE SPEED OF LIGHT.

SAGR.: I have, for instance, seen lead melted instantly by means of a concave mirror only three hands in diameter. Hence I think that if the mirror were very large, well polished and of a parabolic figure, it would just as readily and quickly melt any other metal, seeing that the small mirror, which was not well polished and had only a spherical shape, was able so energetically to melt lead and burn every combustible substance. . . . But now, with regard to the surprising effect of solar rays in melting metals, must we believe that such a furious action is devoid of motion, or that it is accompanied by the most rapid of motions?

SALV.: We observe that other combustions and resolutions are accompanied by motion, and that, the most rapid; note the action of lightning and of power as used in mines and petards; note also how the charcoal flame, mixed as it is with heavy and impure vapors, increases its power to liquify metals whenever quickened by a pair of bellows. Hence I do not understand how the action of light, although very pure, can be devoid of motion and that of the swiftest type.

SAGR.: But of what kind and how great must we consider this speed of light to be? Is it instantaneous or momentary, or does it like other motions require time? Can we not decide this by experiment?

SIMP.: Everyday experience shows that the propagation of light is instantaneous; for when we see a piece of artillery fired, at great distance, the flash reaches our eyes without lapse of time; but the sound reaches the ear only after a noticeable interval.

SAGR.: Well, Simplicio, the only thing I am able to infer from this familiar bit of experience is that sound, in reaching our ear, travels more slowly than light; it does not inform me whether the coming of the light is instantaneous or whether, although extremely rapid, it still occupies time. An observation of this kind tells us nothing more than one in which it is claimed that "As soon as the sun reaches the horizon its light reaches our eyes"; but who will assure me that these rays had not reached this limit earlier than they reached our vision?

SALV.: The small conclusiveness of these and other familiar observations once led me to devise a method by which one might accurately ascertain whether illumination, i.e., the propagation of light, is really instantaneous. The fact that the speed of sound is as high as it is, assures us that the motion of light cannot fail to be extraordinarily swift. The experiment which I devised was as follows:

Let each of two persons take a light contained in a lantern, or other receptacle, such that by the interposition of the hand, the one can shut off or admit the light to the vision of the other. Next let them stand opposite each other at a distance of a few cubits and practice until they acquire such skill in uncovering and occulting their lights that the instant one sees the light of his companion he will uncover his own. After a few trials the response will be so prompt that without sensible error the uncovering of one light is immediately followed by the uncovering of the other, so that as soon as one exposes his light he will instantly see that of the other. Having acquired skill at this short distance let the two experimenters, equipped as before, take up positions separated by a distance of two or three miles and let them perform the same experiment at night, noting carefully whether the exposures and occultations occur in the same manner as at short distances; if they do, we may safely conclude that the propagation of light is instantaneous; but if time is required at a distance of three miles which, considering the going of one light and the coming of the other, really amounts to six, then the delay ought to be easily observable. If the experiment is to be made at still greater distances, say eight or ten miles, telescopes may be employed, each observer adjusting one for himself at the place where he is to make the experiment at night; then although the lights are not large and are therefore invisible to the naked eye at so great a distance, they can readily be covered and uncovered, since by aid of the telescopes, once adjusted and fixed, they will become easily visible.

SAGR.: This experiment strikes me as a clever and reliable invention. But tell us what you conclude from the results.

SALV.: In fact I have tried the experiment only at a short distance, less than a mile, from which I have not been able to ascertain with certainty whether the appearance of the opposite light was instantaneous or not; but if not instantaneous it is extraordinarily rapid—I should call it momentary; and for the present I should compare it to motion which we see in the lightning flash between clouds eight or ten miles distant from us. We see the beginning of this light—I might say that its head and source—located at a particular place among the clouds; but it immediately spreads to the surrounding ones, which seems to be an argument that at least some time is required for propagation; for if the illumination were instantaneous and not gradual, we should not be able to distinguish its origin—its center, so to speak—from its outlying portions. . . .

VACUUM AND FALLING BODIES

SAGR.: . . . As to the vacua I should like to hear a thorough discussion of Aristotle's demonstration in which he opposes them, and what you, Salviati, have to say in reply. I beg you, Simplicio, that you give us the precise proof of the Philosopher and that you, Salviati, give us the reply.

SIMP.: So far as I remember, Aristotle inveighs against the ancient view that a vacuum is a necessary prerequisite for motion and that the latter could not occur without the former. In opposition to this view Aristotle shows that it is precisely the phenomenon of motion, as we shall see, which renders untenable the idea of a vacuum. His method is to divide the argument into two parts. He first supposes bodies of different weights to move in the same medium; then supposes, one and the same body to move in different media. In the first case, he supposes bodies of different weight to move in one and the same medium with different speeds which stand to one another in the same ratio as the weights; so that, for example, a body which is ten times as heavy as another will move ten times as rapidly as the other. In the second case he assumes that the speeds of one and the same body moving in different media are in inverse ratio to the densities of these media; thus, for instance, if the density of water were ten times that of air, the speed in air would be ten times greater than in water. From this second supposition, he shows that, since the tenuity of a vacuum differs infinitely from that of any medium filled with matter however rare, any body which moves in a plenum through a certain space in a certain time ought to move through a vacuum instantaneously; but instantaneous motion is an impossibility; it is therefore impossible that a vacuum should be produced by motion.

SALV.:The argument is, as you see, *ad hominem,* that is, it is directed against those who thought the vacuum a prerequisite for motion. Now if I admit the argument to be conclusive and concede also that motion cannot take place in a vacuum, the assumption of a vacuum considered absolutely and not with reference to motion, is not thereby invalidated. But to tell you what the ancients might possibly have replied and in order to better understand just how conclusive Aristotle's demonstration is, we may, in my opinion, deny both of his assumptions. And as to the first, I greatly doubt that Aristotle ever tested by experiment whether it be true that two stones, one weighing ten times as much as the other, if allowed to fall, at the same instant, from a height of, say, 100 cubits, would so differ in speed that when the heavier had reached the ground, the other would not have fallen more than 10 cubits.

SIMP.: His language would seem to indicate that he had tried the experiment, because he says: *We see the heavier;* now the word *see* shows that he had made the experiment.

SAGR.: But I, Simplicio, who have made the test can assure you that a cannon ball weighing one or two hundred pounds, or even more, will not reach the ground by as much as a span ahead of a musket ball weighing only half a pound, provided both are dropped from a height of 200 cubits.

SALV.: But, even without further experiment, it is possible to prove clearly, by means of a short and conclusive argument, that a heavier body does not move more rapidly than a lighter one provided both bodies are of the same material and in short such as those mentioned by Aristotle. But tell me, Simplicio, whether you admit that each falling body acquires a definite speed fixed by nature, a velocity which cannot be increased or diminished except by the use of force or resistance.

SIMP.: There can be no doubt but that one and the same body moving in a single medium has a fixed velocity which is determined by nature and which cannot be increased except by the addition of momentum or diminished except by some resistance which retards it.

SALV.: If then we take two bodies whose natural speeds are different, it is clear that on uniting the two, the more rapid one will be partly retarded by the slower, and the slower will be somewhat hastened by the swifter. Do you not agree with me in this opinion?

SIMP.: You are unquestionably right.

SALV.: But if this is true, and if a large stone moves with a speed of, say, eight while a smaller moves with a speed of four, then when they are united, the system will move with a speed less than eight; but the two stones when tied together make a stone larger than that which before moved with a speed of eight. Hence the heavier body moves with less speed than the lighter; an effect which is contrary to your supposition. Thus you see how, from your assumption that the heavier body moves more rapidly than the lighter one, I infer that the heavier body moves more slowly.

SIMP.: I am all at sea because it appears to me that the smaller stone when added to the larger increases its weight and by adding weight I do not see how it can fail to increase its speed or, at least, not to diminish it.

SALV.: Here again you are in error, Simplicio, because it is not true that the smaller stone adds weight to the larger.

SIMP.: This is, indeed, quite beyond my comprehension.

SALV.: It will not be beyond you when I have once shown you the mistake under which you are laboring. Note that it is necessary to distinguish between heavy bodies in motion and the same bodies at rest. A large stone placed in a balance not only acquires additional weight by having another stone placed upon it, but even by the addition of a handful of hemp its weight is augmented six to ten ounces according to the quantity of hemp. But if you tie the hemp to the stone and allow them to fall freely from some height, do you believe that the hemp will press down upon the stone and thus accelerate its motion or do you think the motion will be retarded by a partial upward pressure? One always feels the pressure upon his shoulders when he prevents the motion of a load resting upon him; but if one descends just as rapidly as the load would fall how can it gravitate or press upon him? Do you not see that this would be the same as trying to strike a man with a lance when he is running away from you with a speed which is equal to, or even greater, than that with which you are following him? You must therefore conclude that, during free and natural fall, the small stone does not press upon the larger and consequently does not increase its weight as it does when at rest.

SIMP.: But what if we should place the larger stone upon the smaller?

SALV.: Its weight would be increased if the larger stone moved more rapidly; but we have already concluded that when the small stone moves more slowly it retards to some extent the speed of the larger, so that the combination of the two, which is a heavier body than the larger of the two stones, would move less rapidly, a conclusion which is contrary to your hypothesis. We infer therefore that large and small bodies move with the same speed provided they are of the same specific gravity.

SIMP.: Your discussion is really admirable; yet I do not find it easy to believe that a bird-shot falls as swiftly as a cannon ball.

SALV.: Why not say a grain of sand as rapidly as a grindstone? But, Simplicio, I trust you will not follow the example of many others who divert the discussion from its main intent and fasten upon some statement of mine which lacks a hair's-breadth of the truth and, under this hair, hide the fault of another which is as big as a ship's cable. Aristotle says that "an iron ball of one hundred pounds falling from a height of one hundred cubits reaches the ground before a one-pound ball has fallen a single cubit." I say that they arrive at the same time. You find, on making the experiment, that the larger outstrips the smaller by two finger-breadths, that is, when the larger has reached the ground, the other is short of it by two finger-breadths; now you would not hide behind these two fingers the ninety-nine cubits of Aristotle, nor would you mention my small error and at the same time pass over in silence his very large one. Aristotle declares that bodies of different weights, in the same medium, travel (insofar as their motion depends upon gravity) with speeds which are proportional to their weights; this he illustrates by use of bodies in which it is possible to perceive the pure and unadulterated effect of gravity, eliminating other considerations, for example, figure as being of small importance, influences which are greatly dependent upon the medium which modifies the single effect of gravity alone. Thus we observe that gold, the densest of all substances, when beaten out into a very thin leaf, goes floating through the air; the same thing happens with stone when ground into a very fine powder. But if you wish to maintain the general proposition you will have to show that the same ratio of speeds is preserved in the case of all heavy bodies, and that a stone of twenty pounds moves ten times as rapidly as one of two; but I claim that this is false and that, if they fall from a height of fifty or a hundred cubits, they will reach the earth at the same moment.

SIMP.: Perhaps the result would be different if the fall took place not from a few cubits but from some thousands of cubits.

SALV.:If this were what Aristotle meant you would burden him with another error which would amount to a falsehood; because, since there is no such sheer height available on earth, it is clear that Aristotle could not have made the experiment; yet he wishes to give us the impression of his having performed it when he speaks of such an effect as one which we see.

SIMP.:In fact Aristotle does not employ this principle, but uses the other one which is not, I believe, subject to these same difficulties.

SALV.: But the one is as false as the other; and I am surprised that you yourself do not see the fallacy and that you do not perceive that if it were true that, in media of different densities and different resistances, such as water and air, one and the same body moved in air more rapidly than in water, in proportion as the density of water is greater than that of air, then it would follow that any body which falls through air ought also to fall through water. But this conclusion is false inasmuch as many bodies which descend in air not only do not descend in water, but actually rise.

SIMP.: I do not understand the necessity of your inference; and in addition I will say that Aristotle discusses only those bodies which fall in both media, not those which fall in air but rise in water.

SALV.: The arguments which you advance for the Philosopher are such as he himself would have certainly avoided so as not to aggravate his first mistake. But tell me now whether the density of the water, or whatever it may be that retards the motion, bears a definite ratio to the density of air which is less retardative; and if so fix a value for it at your pleasure.

SIMP.: Such a ratio does exist; let us assume it to be ten; then, for a body which falls in both these media, the speed in water will be ten times slower than in air.

SALV.: I shall now take one of those bodies which fall in air but not in water, say a wooden ball, and I shall ask you to assign to it any speed you please for its descent through air.

SIMP.: Let us suppose it moves with a speed of twenty.

SALV.: Very well. Then it is clear that this speed bears to some smaller speed the same ratio as the density of water bears to that of air; and the value of this smaller speed is two. So that really if we follow exactly the assumption of Aristotle we ought to infer that the wooden ball which falls in air, a substance ten times less-resisting than water, with a speed of twenty would fall in water with a speed of two, instead of coming to the surface from the bottom as it does; unless perhaps you wish to reply, which I do not believe you will, that the rising of the wood through the water is the same as its falling with a speed of two. But since the wooden ball does not go to the bottom, I think you will agree with me that we can find a ball of another material, not wood, which does fall in water with a speed of two.

Two New Sciences

SIMP.: Undoubtedly we can; but it must be of a substance considerably heavier than wood.

SALV.: That is it exactly. But if this second ball falls in water with a speed of two, what will be its speed of descent in air? If you hold to the rule of Aristotle you must reply that it will move at the rate of twenty; but twenty is the speed which you yourself have already assigned to the wooden ball; hence this and the other heavier ball will each move through air with the same speed. But now how does the Philosopher harmonize this result with his other, namely, that bodies of different weight move through the same medium with different speeds—speeds which are proportional to their weights? But without going into the matter more deeply, how have these common and obvious properties escaped your notice? Have you not observed that two bodies which fall in water, one with a speed a hundred times as great as that of the other, will fall in air with speeds so nearly equal that one will not surpass the other as much as one hundredth part? Thus, for example, an egg made of marble will descend in water one hundred times more rapidly than a hen's egg, while in air falling from a height of twenty cubits the one will fall short of the other by less than four finger-breadths. In short, a heavy body which sinks through ten cubits of water in three hours will traverse ten cubits of air in one or two pulse-beats; and if the heavy body be a ball of lead it will easily traverse the ten cubits of water in less than double the time required for ten cubits of air. And here, I am sure, Simplicio, you find no ground for difference or objection. We conclude, therefore, that the argument does not bear against the existence of a vacuum; but if it did, it would only do away with vacua of considerable size which neither I nor, in my opinion, the ancients ever believed to exist in nature, although they might possibly be produced by force, as may be gathered from various experiments whose description would here occupy too much time. . . .

WEIGHT DOES NOT AFFECT THE SPEED
OF FALLING BODIES

SALV.: The facts set forth by me up to this point and, in particular, the one which shows that difference of weight, even when very great, is without effect in changing the speed of falling bodies, so that as far as weight is concerned they all fall with equal speed: this idea is, I say, so new, and at first glance so remote from fact, that if we do not have the means of making it just as clear as sunlight, it had better not be mentioned; but having once allowed it to pass my lips I must neglect no experiment or argument to establish it.

SAGR.: Not only this but also many other of your views are so far removed from the commonly accepted opinions and doctrines that if you were to publish them you would stir up a large number of antagonists; for human nature is such that men do not look with favor upon discoveries—either of truth or fallacy—in their own field, when made by others than themselves. They call him an innovator of doctrine, an unpleasant title, by which they hope to cut those knots which they cannot untie, and by subterranean mines they seek to destroy structures which patient artisans have built with customary tools. But as for ourselves who have no such thoughts, the experiments and arguments which you have thus far adduced are fully satisfactory; however if you have any experiments which are more direct or any arguments which are more convincing we will hear them with pleasure.

SALV.: The experiment made to ascertain whether two bodies, differing greatly in weight will fall from a given height with the same speed offers some difficulty; because, if the height is considerable, the retarding effect of the medium, which must be penetrated and thrust aside by the falling body, will be greater in the case of the small momentum of the very light body than in the case of the great force of the heavy body; so that, in a long distance, the light body will be left behind; if the height be small, one may well doubt whether there is any difference; and if there be a difference it will be inappreciable.

It occurred to me therefore to repeat many times the fall through a small height in such a way that I might accumulate all those small intervals of time that elapse between the arrival of the heavy and light bodies respectively at their common terminus, so that this sum makes an interval of time which is not only observable, but easily observable. In order to employ the slowest speeds possible and thus reduce the change which the resisting medium produces upon the simple effect of gravity, it occurred to me to allow the bodies to fall along a plane slightly inclined to the horizontal. For in such a plane, just as well as in a vertical plane, one may discover how bodies of different weight behave: and besides this, I also wished to rid myself of the resistance which might arise from contact of the moving body with the aforesaid inclined plane. Accordingly I took two balls, one of lead and one of cork, the former more than a hundred times heavier than the latter, and suspended them by means of two equal fine threads, each four or five cubits long. Pulling each ball aside from the perpendicular, I let them go at the same instant, and they, falling along the circumferences of circles having these equal strings for semi-diameters, passed beyond the perpendicular and returned along the same path. This free vibration repeated a hundred times showed clearly that the heavy body maintains so nearly the period of the light body that neither in a hundred swings nor even in a thousand will the former anticipate the latter by as much as a single moment, so perfectly do they keep step. We can also observe the effect of the medium which, by the resistance which it offers to motion, diminishes the vibration of the cork more than that of the lead, but without altering the frequency of either; even when the arc traversed by the cork did not exceed five or six degrees while that of the lead was fifty or sixty, the swings were performed in equal times.

SIMP.: If this be so, why is not the speed of the lead greater than that of the cork, seeing that the former traverses sixty degrees in the same interval in which the latter covers scarcely six?

SALV.: But what would you say, Simplicio, if both covered their paths in the same time when the cork, drawn aside through thirty degrees, traverses an arc of sixty, while the lead pulled aside only two degrees traverses an arc of four? Would not then the cork be proportionately swifter? And yet such is the experimental fact. But observe this: having pulled aside the pendulum of lead, say through an arc of fifty degrees, and set it free, it swings beyond the perpendicular almost fifty degrees, thus describing an arc of nearly one hundred degrees; on the return swing it describes a little smaller arc; and after a large number of such vibrations it finally comes to rest. Each vibration, whether of ninety, fifty, twenty, ten, or four degrees occupies the same time: accordingly the speed of the moving body keeps on diminishing since in equal intervals of time, it traverses arcs which grow smaller and smaller.

Precisely the same things happen with the pendulum of cork, suspended by a string of equal length, except that a smaller number of vibrations is required to bring it to rest, since on account of its lightness it is less able to overcome the resistance of the air; nevertheless the vibrations, whether large or small, are all performed in time-intervals which are not only equal among themselves, but also equal to the period of the lead pendulum. Hence it is true that, if while the lead is traversing an arc of fifty degrees the cork covers one of only ten, the cork moves more slowly than the lead; but on the other hand it is also true that the cork may cover an arc of fifty while the lead passes over one of only ten or six; thus, at different times, we have now the cork, now the lead, moving more rapidly. But if these same bodies traverse equal arcs in equal times we may rest assured that their speeds are equal.

SIMP.:I hesitate to admit the conclusiveness of this argument because of the confusion which arises from your making both bodies move now rapidly, now slowly and now very slowly, which leaves me in doubt as to whether their velocities are always equal.

SAGR.: Allow me, if you please, Salviati, to say just a few words. Now tell me, Simplicio, whether you admit that one can say with certainty that the speeds of the cork and the lead are equal whenever both, starting from rest at the same moment and descending the same slopes, always traverse equal spaces in equal times?

SIMP.: This can neither be doubted nor gainsaid.

SAGR.: Now it happens, in the case of the pendulums, that each of them traverses now an arc of sixty degrees, now one of fifty, or thirty or ten or eight or four or two, etc.; and when they both swing through an arc of sixty degrees they do so in equal intervals of time; the same thing happens when the arc is fifty degrees or thirty or ten or any other number; and therefore we conclude that the speed of the lead in an arc of sixty degrees is equal to the speed of the cork when the latter also swings through an arc of sixty degrees; in the case of a fifty-degree arc these speeds are also equal to each other; so also in the case of other arcs. But this is not saying that the speed which occurs in an arc of sixty is the same as that which occurs in an arc of fifty; nor is the speed in an arc of fifty equal to that in one of thirty, etc.; but the smaller the arcs, the smaller the speeds; the fact observed is that one and the same moving body requires the same time for traversing a large arc of sixty degrees as for a small arc of fifty or even a very small arc of ten; all these arcs, indeed, are covered in the same interval of time. It is true therefore that the lead and the cork each diminish their speed in proportion as their arcs diminish; but this does not contradict the fact that they maintain equal speeds in equal arcs.

My reason for saying these things has been rather because

I wanted to learn whether I had correctly understood Salviati, than because I thought Simplicio had any need of a clearer explanation than that given by Salviati, which like everything else of his is extremely lucid, so lucid, indeed, that when he solves questions which are difficult not merely in appearance, but in reality and in fact, he does so with reasons, observations and experiments which are common and familiar to everyone.

In this manner he has, as I have learned from various sources, given occasion to a highly esteemed professor for undervaluing his discoveries on the ground that they are commonplace, and established upon a mean and vulgar basis; as if it were not a most admirable and praiseworthy feature of demonstrative science that it springs from and grows out of principles well known, understood and conceded by all.

But let us continue with this light diet; and if Simplicio is satisfied to understand and admit that the gravity inherent in various falling bodies has nothing to do with the difference of speed observed among them, and that all bodies, insofar as their speeds depend upon it, would move with the same velocity, pray tell us, Salviati, how you explain the appreciable and evident inequality of motion; please reply also to the objection urged by Simplicio—an objection in which I concur—namely, that a cannon ball falls more rapidly than a bird-shot. From my point of view, one might expect the difference of speed to be small in the case of bodies of the same substance moving through any single medium, whereas the larger ones will descend, during a single pulse-beat, a distance which the smaller ones will not traverse in an hour, or in four, or even in twenty hours; as for instance in the case of stones and fine sand and especially that very fine sand which produces muddy water and which in many hours will not fall through as much as two cubits, a distance which stones not much larger will traverse in a single pulse-beat.

SALV.: The action of the medium in producing a greater retardation upon those bodies which have a less specific gravity has already been explained by showing that they experience a diminution of weight. But to explain how one and the same medium produces such different retardations in bodies which are made of the same material and have the same shape, but differ only in size, requires a discussion more clever than that by which one explains how a more expanded shape or an opposing motion of the medium retards the speed of the moving body. The solution of the present problem lies, I think, in the roughness and porosity which are generally and almost necessarily found in the surfaces of solid bodies. When the body is in motion these rough places strike the air or other ambient medium. The evidence for this is found in the humming which accompanies the rapid motion of a body through air, even when that body is as round as possible. One hears not only humming, but also hissing and whistling, whenever there is any appreciable cavity or elevation upon the body. We observe also that a round solid body rotating in a lathe produces a current of air. But what more do we need? When a top spins on the ground at its greatest speed do we not hear a distinct buzzing of high pitch? This sibilant note diminishes in pitch as the speed of rotation slackens, which is evidence that these small rugosites on the surface meet resistance in the air. There can be no doubt, therefore, that in the motion of falling bodies these rugosities strike the surrounding fluid and retard the speed; and this they do so much the more in proportion as the surface is larger,. which is the case of small bodies as compared with greater.

SIMP.: Stop a moment please, I am getting confused. For although I understand and admit that friction of the medium upon the surface of the body retards its motion and that, if other things are the same, the larger surface suffers greater retardation, I do not see on what ground you say that the surface of the smaller body is larger. Besides if, as you say, the larger surface suffers greater retardation the larger solid should move more slowly, which is not the fact. But this objection can be easily met by saying that, although the larger body has a larger surface, it has also a greater weight, in comparison with which the resistance of the larger surface is no more than the resistance of the small surface in comparison with its smaller weight; so that the speed of the larger solid does not become less. I therefore see no reason for expecting any difference of speed so long as the driving weight diminishes in the same proportion as the retarding power of the surface.

SALV.: I shall answer all your objections at once. You will admit, of course, Simplicio, that if one takes two equal bodies, of the same material and same figure, bodies which would therefore fall with equal speeds, and if he diminishes the weight of one of them in the same proportion as its surface (maintaining the similarity of shape) he would not thereby diminish the speed of this body.

SIMP.: This inference seems to be in harmony with your theory which states that the weight of a body has no effect in either accelerating or retarding its motion.

SALV.: I quite agree with you in this opinion from which it appears to follow that, if the weight of a body is diminished in greater proportion than its surface, the motion is retarded to a certain extent; and this retardation is greater and greater in proportion as the diminution of weight exceeds that of the surface.

SIMP.: This I admit without hesitation.

SALV.: Now you must know, Simplicio, that it is not possible to diminish the surface of a solid body in the same ratio as the weight, and at the same time maintain similarity of figure. For since it is clear that in the case of a diminishing solid the weight grows less in proportion to the volume, and since the volume always diminishes more rapidly than the surface, when the same shape is maintained, the weight must therefore diminish more rapidly than the surface. But geometry teaches us that, in the case of similar solids, the ratio of two volumes is greater than the ratio of their surfaces; which, for the sake of better understanding, I shall illustrate by a particular case.

Take, for example, a cube two inches on a side so that each face has an area of four square inches and the total area, i.e., the sum of the six faces, amounts to twenty-four square inches; now imagine this cube to be sawed through three times so as to divide it into eight smaller cubes, each one inch on the side, each face one inch square, and the total surface of each cube six square inches instead of twenty-four as in the case of the larger cube. It is evident therefore that the surface of the little cube is only one-fourth that of the larger, namely, the ratio of six to twenty-four; but the volume of the solid cube itself is only one-eighth; the volume, and hence also the weight, diminishes therefore much more rapidly than the surface. If we again divide the little cube into eight others we shall have, for the total surface of one of these, one and one-half square inches, which is one-sixteenth of the surface of the original cube; but its volume is only one-sixty-fourth part. Thus, by two divisions, you see that the volume is diminished four times as much as the surface. And, if the subdivision be continued until the original solid be reduced to a fine powder, we shall find that the weight of one of these smallest particles has diminished hundreds and hundreds of times as much as its surface. And this which I have illustrated in the case of cubes holds also in the case of all similar solids, where the volumes stand in sesquialteral ratio to their surfaces. Observe then how much greater the resistance, arising from contact of the surface of the moving body with the medium, in the case of small bodies than in the case of large; and when one considers that the rugosities on the very small surfaces of fine dust particles are perhaps no smaller than those on the surfaces of larger solids which have been carefully polished, he will see how important it is that the medium should be very fluid and offer no resistance to being thrust aside, easily yielding to a small force. You see, therefore, Simplicio, that I was not mistaken when, not long ago, I said that the surface of a small solid is comparatively greater than that of a large one.

SIMP.: I am quite convinced; and, believe me, if I were again beginning my studies, I should follow the advice of Plato and start with mathematics, a science which proceeds very cautiously and admits nothing as established until it has been rigidly demonstrated. . . .

THE PENDULUM AND MUSICAL PROPORTIONS

SALV.: We come now to the other questions, relating to pendulums, a subject which may appear to many exceedingly arid, especially to those philosophers who are continually occupied with the more profound questions of nature. Nevertheless, the problem is one which I do not scorn. I am encouraged by the example of Aristotle whom I admire especially because he did not fail to discuss every subject which he thought in any degree worthy of consideration.

Impelled by your queries I may give you some of my ideas concerning certain problems in music, a splendid subject, upon which so many eminent men have written: among these is Aristotle himself who has discussed numerous interesting acoustical questions. Accordingly, if on the basis of some easy and tangible experiments, I shall explain some striking phenomena in the domain of sound, I trust my explanations will meet your approval.

SAGR.: I shall receive them not only gratefully but eagerly. For, although I take pleasure in every kind of musical instrument and have paid considerable attention to harmony, I have never been able fully to understand why some combinations of tones are more pleasing than others, or why certain combinations not only fail to please but are even highly offensive. Then there is the old problem of two stretched strings in unison; when one of them is sounded, the other begins to vibrate and to emit its note; nor do I understand the different ratios of harmony and some other details.

SALV.: Let us see whether we cannot derive from the pendulum a satisfactory solution of all these difficulties. And first, as to the question whether one and the same pendulum really performs its vibrations, large, medium, and small, all in exactly the same time, I shall rely upon what I have already heard from our Academician. He has clearly shown that the time of descent is the same along all chords, whatever the arcs which subtend them, as well along an arc of 180° (i.e., the whole diameter) as along one of 100°, 60°, 10°, 2°, ½°, or 4'. It is understood, of course, that these arcs all terminate at the lowest point of the circle, where it touches the horizontal plane.

If now we consider descent along arcs instead of their chords then, provided these do not exceed 90°, experiment shows that they are all traversed in equal times; but these times are greater for the chord than for the arc, an effect which is all the more remarkable because at first glance one would think just the opposite to be true. For since the terminal points of the two motions are the same and since the straight line included between these two points is the shortest distance between them, it would seem reasonable that motion along this line should be executed in the shortest time; but this is not the case, for the shortest time—and therefore the most rapid motion—is that employed along the arc of which this straight line is the chord.

As to the times of vibration of bodies suspended by threads of different lengths, they bear to each other the same proportion as the square roots of the lengths of the thread; or one might say the lengths are to each other as the squares of the times; so that if one wishes to make the vibration-time of one pendulum twice that of another, he must make its suspension four times as long. In like manner, if one pendulum has a suspension nine times as long as another, this second pendulum will execute three vibrations during each one of the first; from which it follows that the lengths of the suspending cords bear to each other the [inverse] ratio of the squares of the number of vibrations performed in the same time.

SAGR.: Then, if I understand you correctly, I can easily measure the length of a string whose upper end is attached at any height whatever even if this end were invisible and I could see only the lower extremity. For if I attach to the lower end of this string a rather heavy weight and give it a to-and-fro motion, and if I ask a friend to count a number of its vibrations, while I, during the same time-interval, count the number of vibrations of a pendulum which is exactly one cubit in length, then knowing the number of vibrations which each pendulum makes in the given interval of time one can determine the length of the string. Suppose, for example, that my friend counts 20 vibrations of the long cord during the same time in which I count 240 of my string which is one cubit in length; taking the squares of the two numbers, 20 and 240, namely 400 and 57,600, then, I say, the long string contains 57,600 units of such length that my pendulum will contain 400 of them; and since the length of my string is one cubit, I shall divide 57,600 by 400 and thus obtain 144. Accordingly I shall call the length of the string 144 cubits.

SALV.: Nor will you miss it by as much as a hand's breadth, especially if you observe a large number of vibrations.

SAGR.: You give me frequent occasion to admire the wealth and profusion of nature when, from such common and even trivial phenomena, you derive facts which are not only striking and new but which are often far removed from what we would have imagined. Thousands of times I have observed vibrations especially in churches where lamps, suspended by long cords, had been inadvertently set into motion; but the most which I could infer from these observations was that the view of those who think that such vibrations are maintained by the medium is highly improbable: for, in that case, the air must needs have considerable judgment and little else to do but kill time by pushing to and fro a pendent weight with perfect regularity. But I never dreamed of learning that one and the same body, when suspended from a string a hundred cubits long and pulled aside through an arc of 90° or even 1° or ½°, would employ the same time in passing through the least as through the largest of these arcs; and, indeed, it still strikes me as somewhat unlikely. Now I am waiting to hear how these same simple phenomena can furnish solutions for those acoustical problems—solutions which will be at least partly satisfactory.

SALV.: First of all one must observe that each pendulum has its own time of vibration so definite and determinate that it is not possible to make it move with any other period than that which nature has given it. For let any one take in his hand the cord to which the weight is attached and try, as much as he pleases, to increase or diminish the frequency of its vibrations; it will be time wasted. On the other hand, one can confer motion upon even a heavy pendulum which is at rest by simply blowing against it; by repeating these blasts with a frequency which is the same as that of the pendulum one can impart considerable motion. Suppose that by the first puff we have displaced the pendulum from the vertical by, say, half an inch; then if, after the pendulum has returned and is about to begin the second vibration, we add a second puff, we shall impart additional motion; and so on with other blasts provided they are applied at the right instant, and not when the pendulum is coming toward us since in this case the blast would impede rather than aid the motion. Continuing thus with many impulses we impart to the pendulum such momentum that a greater impulse than that of a single blast will be needed to stop it.

SAGR.: Even as a boy, I observed that one man alone by giving these impulses at the right instant was able to ring a bell so large that when four, or even six, men seized the rope and tried to stop it they were lifted from the ground, all of them together being unable to counterbalance the momentum which a single man, by properly-timed pulls, had given it.

SALV.: Your illustration makes my meaning clear and is quite as well fitted, as what I have just said, to explain the wonderful phenomenon of the strings of the cittern or of the spinet, namely, the fact that a vibrating string will set another string in motion and cause it to sound not only when the latter is in unison but even when it differs from the former by an octave or a fifth. A string which has been struck begins to vibrate and continues the motion as long as one hears the sound; these vibrations cause the immediately surrounding air to vibrate and quiver; then these ripples in the air expand far into space and strike not only all the strings of the same instrument but even those of neighboring instruments. Since that string which is tuned to unison with the one plucked is capable of vibrating with the same frequency, it acquires, at the first impulse, a slight oscillation; after receiving two, three, twenty, or more impulses, delivered at proper intervals, it finally accumulates a vibratory motion equal to that of the plucked string, as is clearly shown by equality of amplitude in their vibrations. This undulation expands through the air and sets into vibration not only strings, but also any other body which happens to have the same period as that of the plucked string. Accordingly if we attach to the side of an instrument small pieces of bristle or other flexible bodies, we shall observe that, when a spinet is sounded, only those pieces respond that have the same period as the string which has been struck; the remaining pieces do not vibrate in response to this string, nor do the former pieces respond to any other tone.

If one bows the base string on a viola rather smartly and brings near it a goblet of fine, thin glass having the same tone as that of the string, this goblet will vibrate and audibly resound. That the undulations of the medium are widely dispersed about the sounding body is evinced by the fact that a glass of water may be made to emit a tone merely by the friction of the finger-tip upon the rim of the glass; for in this water is produced a series of regular waves. The same phenomenon is observed to better advantage by fixing the base of the goblet upon the bottom of a rather large vessel of water filled nearly to the edge of the goblet; for it, as before, we sound the glass by friction of the finger, we shall see ripples spreading with the utmost regularity and with high speed to large distances about the glass. I have often remarked, in thus sounding a rather large glass nearly full of water, that at first the waves are spaced with great uniformity, and when, as sometimes happens, the tone of the glass jumps an octave higher I have noted that at this moment each of the aforesaid waves divides into two; a phenomenon which shows clearly that the ratio involved in the octave is two.

SAGR.: More than once have I observed this same thing, much to my delight and also to my profit. For a long time I have been perplexed about these different harmonies since the explanations hitherto given by those learned in music impress me as not sufficiently conclusive. . . .

SALV.: Seeing that you have derived so much pleasure from these novelties, I must show you a method by which the eye may enjoy the same game as the ear. Suspend three balls of lead, or other heavy material, by means of strings of different length such that while the longest makes two vibrations the shortest will make four and the medium three; this will take place when the longest string measures 16, either in hand breadths or in any other unit, the medium 9 and the shortest 4, all measured in the same unit.

Now pull all these pendulums aside from the perpendicular and release them at the same instant; you will see a curious interplay of the threads passing each other in various manners but such that at the completion of every fourth vibration of the longest pendulum, all three will arrive simultaneously at the same terminus, whence they start over again to repeat the same cycle. This combination of vibrations, when produced on strings is precisely that which yields the interval of the octave and the intermediate fifth. If we employ the same disposition of the apparatus but change the lengths of the threads, always however in such a way that their vibrations correspond to those of agreeable musical intervals, we shall see a different crossing of these threads but always such that, after a definite interval of time and after a definite number of vibrations, all the threads, whether three or four, will reach the same terminus at the same instant, and then begin a repetition of the cycle.

If, however, the vibrations of two or more strings are incommensurable so that they never complete a definite number of vibrations at the same instant, or if commensurable they return only after a long interval of time and after a large number of vibrations, then the eye is confused by the disorderly succession of crossed threads. In like manner the ear is pained by an irregular sequence of air waves which strike the tympanum without any fixed order.

But, gentlemen, whither have we drifted during these many hours lured on by various problems and unexpected digressions? The day is already ended and we have scarcely touched the subject proposed for discussion. Indeed we have deviated so far that I remember only with difficulty our early introduction and the little progress made in the way of hypotheses and principles for use in later demonstrations.

SAGR.: Let us then adjourn for today in order that our minds may find refreshment in sleep and that we may return tomorrow, if so please you, and resume the discussion of the main questions.

SALV.: I shall not fail to be here tomorrow at the same hour, hoping not only to render you service but also to enjoy your company. . . .

SECOND NEW SCIENCE, TREATING OF MOTION

My purpose is to set forth a very new science dealing with a very ancient subject. There is, in nature, perhaps nothing older than motion, concerning which the books written by philosophers are neither few nor small; nevertheless I have discovered by experiment some properties of it which are worth knowing and which have not hitherto been either observed or demonstrated. Some superficial observations have been made, as, for instance, that the free motion of a heavy falling body is continuously accelerated; but to just what extent this acceleration occurs has not yet been announced; for so far as I know, no one has yet pointed out that the distances traversed, during equal intervals of time, by a body falling from rest, stand to one another in the same ratio as the odd numbers beginning with unity.

It has been observed that missiles and projectiles describe a curved path of some sort; however, no one has pointed out the fact that this path is a parabola. But this and other facts, not few in number or less worth knowing, I have succeeded in proving; and what I consider more important, there have been opened up to this vast and most excellent science, of which my work is merely the beginning, ways and means by which other minds more acute than mine will explore its remote corners.

This discussion is divided into three parts; the first part deals with motion which is steady or uniform; the second [presented here] treats of motion as we find it accelerated in nature; the third deals with the so-called violent motions and with projectiles. . . .

NATURALLY ACCELERATED MOTION

The properties belonging to uniform motion have been discussed in the preceding section; but accelerated motion remains to be considered.

And first of all it seems desirable to find and explain a definition best fitting natural phenomena. For anyone may invent an arbitrary type of motion and discuss its properties; thus, for instance, some have imagined helices and conchoids as described by certain motions which are not met with in nature, and have very commendably established the properties which these curves possess in virtue of their definitions; but we have decided to consider the phenomena of bodies falling with an acceleration such as actually occurs in nature and to make this definition of accelerated motion exhibit the essential features of observed accelerated motions. And this, at last, after repeated efforts we trust we have succeeded in doing. In this belief we are confirmed mainly by the consideration that experimental results are seen to agree with and exactly correspond with those properties which have been, one after another, demonstrated by us. Finally, in the investigation of naturally accelerated motion we were led, by hand as it were, in following the habit and custom of nature herself, in all her various other processes, to employ only those means which are most common, simple and easy.

For I think no one believes that swimming or flying can be accomplished in a manner simpler or easier than that instinctively employed by fishes and birds.

When, therefore, I observe a stone initially at rest falling from an elevated position and continually acquiring new increments of speed, why should I not believe that such increases take place in a manner which is exceedingly simple and rather obvious to everybody? If now we examine the matter carefully we find no addition or increment more simple than that which repeats itself always in the same manner. This we readily understand when we consider the intimate relationship between time and motion; for just as uniformity of motion is defined by and conceived through equal times and equal spaces (thus we call a motion uniform when equal distances are traversed during equal time-intervals), so also we may, in a similar manner, through equal time-intervals, conceive additions of speed as taking place without complication; thus we may picture to our mind a motion as uniformly and continuously accelerated when, during any equal intervals of time whatever, equal increments of speed are given to it. Thus if any equal intervals of time whatever have elapsed, counting from the time at which the moving body left its position of rest and began to descend, the amount of speed acquired during the first two time-intervals will be double that acquired during the first time-interval alone; so the amount added during three of these time-intervals will be treble; and that in four, quadruple that of the first time-interval. To put the matter more clearly, if a body were to continue its motion with the same speed which it had acquired during the first time-interval and were to retain this same uniform speed, then its motion would be twice as slow as that which it would have if its velocity had been acquired during *two* time-intervals.

And thus, it seems, we shall not be far wrong if we put the increment of speed as proportional to the increment of time; hence the definition of motion which we are about to discuss may be started as follows: A motion is said to be uniformly accelerated, when starting from rest, it acquires, during equal time-intervals, equal increments of speed.

SAGR.: Although I can offer no rational objection to this or indeed to any other definition, devised by any author whomsoever, since all definitions are arbitrary, I may nevertheless without offense be allowed to doubt whether such a definition as the above, established in an abstract manner, corresponds to and describes that kind of accelerated motion which we meet in nature in the case of freely falling bodies. And since the Author apparently maintains that the motion described in his definition is that of freely falling bodies, I would like to clear my mind of certain difficulties in order that I may later apply myself more earnestly to the propositions and their demonstrations.

SALV.: It is well that you and Simplicio raise these difficulties. They are, I imagine, the same which occurred to me when I first saw this treatise, and which were removed either by discussion with the Author himself, or by turning the matter over in my own mind.

SAGR.: When I think of a heavy body falling from rest, that is, starting with zero speed and gaining speed in proportion to the time from the beginning of the motion; such a motion as would, for instance, in eight beats of the pulse acquire eight degrees of speed; having at the end of the fourth beat acquired four degrees; at the end of the second, two; at the end of the first, one: and since time is divisible without limit, it follows from all these considerations that if the earlier speed of a body is less than its present speed in a constant ratio, then there is no degree of speed however small (or, one may say, no degree of slowness however great) with which we may not find this body travelling after starting from infinite slowness, i.e., from rest. So that if that speed which it had at the end of the fourth beat was such that, if kept uniform, the body would traverse two miles in an hour, and if keeping the speed which it had at the end of the second beat, it would traverse one mile an hour, we must infer that, as the instant of starting is more and more nearly approached, the body moves so slowly that, if it kept on moving at this rate, it would not traverse a mile in an hour, or in a day, or in a year or in a thousand years; indeed, it would not traverse a span in an even greater time; a phenomenon which baffles the imagination, while our senses show us that a heavy falling body suddenly acquires great speed.

SALV.: This is one of the difficulties which I also at the beginning, experienced, but which I shortly afterwards removed; and the removal was effected by the very experiment which creates the difficulty for you. You say the experiment appears to show that immediately after a heavy body stars from rest it acquires a very considerable speed: and I say that the same experiment makes clear the fact that the initial motions of a falling body, no matter how heavy, are very slow and gentle. Place a heavy body upon a yielding material, and leave it there without any pressure except that owing to its own weight; it is clear that if one lifts this body a cubit or two and allows it to fall upon the same material, it will, with this impulse, exert a new and greater pressure than that caused by its mere weight; and this effect is brought about by the [weight of the] falling body together with the velocity acquired during the fall, an effect which will be greater and greater according to the height of the fall, that is according as the velocity of the falling body becomes greater. From the quality and intensity of the blow we are thus enabled accurately to estimate the speed of a falling body. But tell me, gentlemen, is it not true that if a block be allowed to fall upon a stake from a height of four cubits and drives it into the earth, say, four finger-breadths, that coming from a height of two cubits it will drive the stake a much less distance, and from the height of one cubit a still less distance; and finally if the block be lifted only one finger-breadth how much more will it accomplish than if merely laid on top of the stake without percussion? Certainly very little. If it be lifted only the thickness of a leaf, the effect will be altogether imperceptible. And since the effect of the blow depends upon the velocity of this striking body, can any one doubt the motion is very slow and the speed more than small whenever the effect [of the blow] is imperceptible? See now the power of truth; the same experiment which at first glance seemed to show one thing, when more carefully examined, assures us of the contrary.

But without depending upon the above experiment, which is doubtless very conclusive, it seems to me that it ought not to be difficult to establish such a fact by reasoning alone. Imagine a heavy stone held in the air at rest; the support is removed and the stone set free; then since it is heavier than the air it begins to fall, and not with uniform motion but slowly at the beginning and with a continuously accelerated motion. Now since velocity can be increased and diminished without limit, what reason is there to believe that such a moving body starting with infinite slowness, that is, from rest, immediately acquires a speed of ten degrees rather than one of four, or of two, or of one, or of a half, or of a hundredth; or, indeed, of any of the infinite number of small values [of speed]? Pray listen. I hardly think you will refuse to grant that the gain of speed of the stone falling from rest follows the same sequence as the diminution and loss of this same speed when, by some impelling force, the stone is thrown to its former elevation: but even if you do not grant this, I do not see how you can doubt that the ascending stone, diminishing in speed, must before coming to rest pass through every possible degree of slowness.

SIMP.: But if the number of degrees of greater and greater slowness is limitless, they will never be all exhausted, therefore such an ascending heavy body will never reach rest, but will continue to move without limit always at a slower rate; but this is not the observed fact.

SALV.: This would happen, Simplicio, if the moving body were to maintain its speed for any length of time at each degree of velocity; but it merely passes each point without delaying more than an instant: and since each time-interval however small may be divided into an infinite number of instants, these will always be sufficient [in number] to correspond to the infinite degrees of diminished velocity.

That such a heavy rising body does not remain for any length of time at any given degree of velocity is evident from the following: because if, some time-interval having been assigned, the body moves with the same speed in the first instant of that time-interval, it could from this second degree of elevation be in like manner raised through an equal height, just as it was transferred from the first elevation to the second, and by the same reasoning would pass from the second to the third and would finally continue in uniform motion forever.

SAGR.: From these considerations it appears to me that we may obtain a proper solution of the problem discussed by philosophers, namely, what causes the acceleration in the natural motion of heavy bodies? Since, as it seems to me, the force impressed by the agent projecting the body upwards diminishes continuously, this force, so long as it was greater than the contrary force of gravitation, impelled the body upwards; when the two are in equilibrium the body ceases to rise and passes through the state of rest in which the impressed impetus is not destroyed, but only its excess over the weight of the body has been consumed—the excess which caused the body to rise. Then as the diminution of the outside impetus continues, and gravitation gains the upper hand, the fall begins, but slowly at first on account of the opposing impetus, a large portion of which still remains in the body; but as this continues to diminish it also continues to be more and more overcome by gravity, hence the continuous acceleration of motion.

SIMP.: The idea is clever, yet more subtle than sound; for even if the argument were conclusive, it would explain only the case in which a natural motion is preceded by a violent motion, in which there still remains active a portion of the external force; but where there is no such remaining portion and the body starts from an antecedent state of rest, the cogency of the whole argument fails.

SAGR.: I believe that you are mistaken and that this distinction between cases which you make is superfluous or rather nonexistent. But, tell me, cannot a projectile receive from the projector either a large or a small force such as will throw it to a height of a hundred cubits, and even twenty or four or one?

SIMP.: Undoubtedly, yes.

SAGR.: So therefore this impressed force may exceed the resistance of gravity so slightly as to raise it only a fingerbreadth; and finally the force of the projector may be just large enough exactly to balance the resistance of gravity so that the body is not lifted at all but merely sustained. When one holds a stone in his hand does he do anything but give it a force impelling it upwards equal to the power of gravity drawing it downwards? And do you not continuously impress this force upon the stone as long as you hold it in the hand? Does it perhaps diminish with the time during which one holds the stone?

And what does it matter whether this support which prevents the stone from falling is furnished by one's hand or by a table or by a rope from which it hangs? Certainly nothing at all. You must conclude, therefore, Simplicio, that it makes no difference whatever whether the fall of the stone is preceded by a period of rest which is long, short, or instantaneous provided only the fall does not take place so long as the stone is acted upon by a force opposed to its weight and sufficient to hold it at rest.

SALV.: The present does not seem to be the proper time to investigate the cause of the acceleration of natural motion concerning which various opinions have been expressed by various philosophers, some explaining it by attraction to the center, others to repulsion between the very small parts of the body, while still others attribute it to a certain stress in the surrounding medium which closes in behind the falling body and drives it from one of its positions to another. Now, all these fantasies, and others too, ought to be examined; but it is not really worth while. At present it is the purpose of our Author merely to investigate and to demonstrate some of the properties of accelerated motion (whatever the cause of this acceleration may be)—meaning thereby a motion, such that the momentum of its velocity goes on increasing after departure from rest, in simple proportionality to the time, which is the same as saying that in equal time-intervals the body receives equal increments of velocity; and if we find the properties [of accelerated motion] which will be demonstrated later are realized in freely falling and accelerated bodies, we may conclude that the assumed definition includes such a motion of falling bodies and that their speed goes on increasing as the time and the duration of the motion.

SAGR.: So far as I see at present, the definition might have been put a little more clearly perhaps without changing the fundamental idea, namely, uniformly accelerated motion is such that its speed increases in proportion to the space traversed; so that, for example, the speed acquired by a body in falling four cubits would be double that acquired in falling two cubits and this latter speed would be double that acquired in the first cubit. Because there is no doubt but that a heavy body falling from the height of six cubits has, and strikes with, a momentum double that it had at the end of three cubits, triple that which it had at the end of one.

SALV.: It is very comforting to me to have had such a companion in error; and moreover let me tell you that your proposition seems so highly probable that our Author himself admitted, when I advanced this opinion to him, that he had for some time shared the same fallacy. But what most surprised me was to see two propositions so inherently probable that they commanded the assent of everyone to whom they were presented, proven in a few simple words to be not only false, but impossible.

SIMP.: I am one of those who accept the proposition, and believe that a falling body acquires force in its descent, its velocity increasing in proportion to the space, and that the momentum of the falling body is doubled when it falls from a doubled height; these propositions, it appears to me, ought to be conceded without hesitation or controversy.

SALV.: And yet they are as false and impossible as that motion should be completed instantaneously; and here is a very clear demonstration of it. If the velocities are in proportion to the spaces traversed, or to be traversed, then these spaces are traversed in equal intervals of time; if, therefore, the velocity with which the falling body traverses a space of eight feet were double that with which it covered the first four feet (just as the one distance is double the other) then the time-intervals required for these passages would be equal. But for one and the same body to fall eight feet and four feet in the same time is possible only in the case of instantaneous [discontinuous] motion; but observation shows us that the motion of a falling body occupies time, and less of it in covering a distance of four feet than of eight feet; therefore it is not true that its velocity increases in proportion to the space.

The falsity of the other proposition may be shown with equal clearness. For if we consider a single striking body the difference of momentum in its blows can depend only upon difference of velocity; for if the striking body falling from a double height were to deliver a blow of double momentum, it would be necessary for this body to strike with a doubled velocity; but with this doubled speed it would traverse a doubled space in the same time-interval; observation, however, shows that the time required for fall from the greater height is longer.

SAGR.: You present these recondite matters with too much evidence and ease; this great facility makes them less appreciated than they would be had they been presented in a more abstruse manner. For, in my opinion, people esteem more lightly that knowledge which they acquire with so little labor than that acquired through long and obscure discussion.

SALV.: If those who demonstrate with brevity and clearness the fallacy of many popular beliefs were treated with contempt instead of gratitude the injury would be quite bearable; but on the other hand it is very unpleasant and annoying to see men, who claim to be peers of anyone in a certain field of study, take for granted certain conclusions which later are quickly and easily shown by another to be false. I do not describe such a feeling as one of envy, which usually degenerates into hatred and anger against those who discover such fallacies: I would call it a strong desire to maintain old errors, rather than accept newly discovered truths.

CHAPTER 4

The Judgment of Posterity

THE READINGS in this chapter illustrate, historically, the various estimates of the achievement of Galileo that serious men in subsequent ages have entertained down to our own time. No attempt has been made to sample the infinite variety of impressions made by the first publication of Galileo's original telescope observations of the heavens. That history has frequently been told, most interestingly, for English readers, by Professor Marjorie Nicolson in her *Voyages to the Moon, The Breaking of the Circle,* and numerous other writings on the subject.

Neither has any attempt been made to sample the enormous quantity of polemical writing that was called forth by Galileo's literary involvement in the controversy over the Copernican hypothesis and related discussions on the rights and proper methodology of natural science. The first two readings presented here, taken from the writings of Paolo Frisi and Thomas Henri Martin and translated for the first time into English, very adequately summarize and illustrate by numerous quotations the seventeenth- and eighteenth-century estimates of the work of Galileo as a perfecter of the scientific method and as a defender of the rights of science.

The third reading, drawn from the well-known work of Andrew Dickson White, *A History of the Warfare of Science with Theology in Christendom,* is a classic statement of what remains to this day the popularly accepted impression of the significance of Galileo's stand against ecclesiastical dictation in the sphere of science.

The final four readings, from the works of Einstein, Morris R. Cohen, Philipp Frank, and Pierre Duhem (not heretofore translated into English) serve, as was noted in the Introduction, to project the achievement of Galileo against the great world-picture being shaped by twentieth-century science.

from

The Philosophy of Galileo*

by

PAOLO FRISI

[The editors of the 1782 edition of the *Encyclopédie* attributed the article on the *Philosophie de Galilée*, which is here translated, to Paolo Frisi, a celebrated mathematician and physicist of the eighteenth century, professor at the Universities of Pisa and Milan, who was especially distinguished, as the *Encyclopedia Britannica* notes, for his work in hydraulics. But the essay as it appears in the *Encyclopédie* is more than likely a very free adaptation of an original the source of which is not precisely indicated by the editors. Thomas Martin, himself a distinguished scientist and historian of science (see the following selection), has noted, with reference to this article, that Galileo had originally received but little sympathy from the French Encyclopedists and their followers, no doubt because, to please them "he would have had to attack Christianity or at least let himself be burned by the Inquisition as a heretic." In the *Encyclopédie* of Diderot and d'Alembert, Martin noted, "Galileo is hardly mentioned. . . . Only in the supplement to the first edition, in 1775, is there inserted a French translation of the *Eulogy of Galileo* which had just been published in Italian by the Barnabite scientist Frisi. The merits of Galileo are rather well appreciated there; but Frisi is silent about the persecution that he underwent."]

* Translated by James Brophy from the *Encyclopédie, ou dictionnaire Raissoné Des Sciences, Des Arts et Des Métiers,* edited by Diderot and D'Alembert, Lausanne and Berne, 1782. Copyright © 1961 by James Brophy.

In the preface of the proceedings of the Academy of Dijon, one reads a very favorable judgment on the discoveries and on the merit of Galileo. One also reads there that while Francis Bacon was showing England the path to truth, Galileo in Italy already had made great progress in the same direction; that the same Galileo was clairvoyant enough to have discovered the laws of the fall of heavy bodies, laws which, since generalized by Newton, have explained for us the great system of the universe; that he gained by his marvelous instruments a new world for science; that the heavens before his eyes seemed to expand, and the earth people itself with new inhabitants; that Galileo, not content simply with the glory of having made new discoveries, added to it the glory of deriving from them the greatest advantages for the human race; and that after having observed the satellites of Jupiter for twenty-seven years, he used the tables of their movements to determine the longitudes, and to perfect geography and navigation; that his experiments on the weight of air gave birth to a whole new physics, which led Toricelli to explain atmospheric pressure and the suspension of mercury in barometers; that his observations on the movement of the pendulum directed the astronomers and physicists in the process of measuring time with precision, to control the variation of weights in different climates, and to deduce the true shape of the earth; and one concludes that Galileo has discovered much, and has acquired obvious rights over the discoveries of others.

To what the academicians of Dijon have said, one could add the testimony of many Italian authors, who have given the highest praise to Galileo. In Holland, Hugo Grotius says that his works surpass human power; Huygens calls him "a very great man." Leibnitz in Germany, and Jean Bernoulli recognize him "as the profoundest mind of his time," and Kepler writes that he mounts "on the highest walls of the universe," and that his discoveries are things complete from beginning to ending. Newton, in England, several times cites the theorems and discoveries of Galileo. Keill has also written that Galileo, with the aid of geometry, penetrates the most hidden secret of nature and creates a new understanding of motion; and Maclaurin greatly praises the services that he has rendered us with the aid of his telescope, and also the clear and geometric method, with which he explained to us the theory of heavy bodies that fall, or that are projected in some direction. David Hume, in his appendix to the history of James I, makes a very close comparison of Francis Bacon and Galileo. He says that Bacon was inferior to Galileo, his contemporary, and perhaps even to Kepler; that Bacon sees only the road that is shown where Galileo advances with great strides; that the first did not know geometry; that the second knew it perfectly, as well as natural philosophy; that the first scorned the Copernican system, which the second had established by proofs drawn from reason and good sense; that the style of the first was heavy, and that of the second pleasant and brilliant, although at times prolix. The English historian remarks agreeably, that Italy did not, perhaps, make of Galileo the case he merited, simply because of the number of illustrious men who then flourished there.

Galileo Galilei was born in Pisa in 1564, and became a teacher of mathematics there in 1589; three years later he taught at Padua: in 1610 he was made the mathematician of the Grand Duke Ferdinand II, and returned to Tuscany, where he died in 1640 in the town of Accetri, near Florence. He was born in the year that Michelangelo Buonarrotti died at Rome, and he died in the year that Isaac Newton was born in England. In 1583, as Magalotti in his *Essays on the Academy of Cimento,* and Viviani attest, being seated in the cathedral at Pisa, he observed that a lamp set in motion made oscillations that were perceptibly equal in time, although the arcs described were perceptibly unequal to each other. This important observation was carried so far by Galileo, that he conceived of using a pendulum to measure time exactly, and, in his old age, applied it to the clock. Becker, in a dissertation on the measurement of time, attests having heard the Count Magalotti say that Galileo had had the first pendulum clock made at Florence by Marc Treffler, the Grand Duke's clock-maker; yet the same Magalotti in his *Essays on the Academy of Cimento* says that while it was indeed Galileo who conceived of using the pendulum in a clock, it was his son Vincent who, in 1649, put it into practice. We have, however, Galileo's letters to Beaugrand, and those of Realio and d'Ortensius, which, taken with what Viviani says, make it appear beyond any doubt that it was Galileo who added the pendulum to the clock. Elio Diodati, in 1637, sent to the father of the celebrated Huygens the description of the pendulum-clock made by Galileo. Becker adds that a model of it was even sent to Holland. All this suffices to answer Huygens, Musschenbroeck and many others who would like to deprive Italy of glory for this brilliant invention. Huygens invented a pendulum that made its oscillations in the arcs of a cycloid. The invention is very ingenious; and the geometric theory which the inventor gives of it is one of the most brilliant achievements of geometry; but from the point of view of practical convenience the cycloidal pendulum was soon abandoned, and we now use pendulums that move in small circular arcs, such as Galileo originally invented.

While he was teaching at Pisa, he undertook diverse public experiments on the fall of heavy bodies, and he demonstrated for all to see that wood, metals, and other bodies, although of different weights, fall in the same length of time, and with an equal speed, from the same height. He drew from this the important theorem that the absolute gravity of bodies is proportional to the quantity of their matter. In the year 1597, he invented at Padua his proportional compass, which is and always will be a very useful instrument. He was the first who invented the thermometer, and found the way of increasing the force of the magnet 180 times; and having heard, in 1609, that a Hollander had made an optical instrument which brought objects closer, he immediately guessed its construction: he made a copy of it the following day, and six days later took one to Venice that magnified the diameter of objects thirty times. He himself showed in his essay by what simple reasonings, or to put it better, by what simple experiments he had accomplished it. He saw quickly enough that objects could not be enlarged or made more clearly visible with the use of one, or several plane pieces of glass, nor with a concave lens that diminished things, nor with a convex lens that enlarges and blurs them. His task was thus reduced to discovering what a convex and a concave lens used together would produce, and he saw that the result confirmed his idea. Since then telescopes have been made that magnify more or cover a wider field, by using two convex lenses, and other combinations of glass, but these in no way detract from Galileo's theory.

Several authors have found the traces of this discovery in the works of Roger Bacon and Gian Battista Porta, and have attributed to them the invention of the telescope. But the celebrated Robert Smith, in his *Treatise on Optics,* after having examined all the fragments of Roger Bacon, made it evident that this man, whom M. de Voltaire had already called "a gold piece encrusted with the dirt of his century," not only did not have the idea of the telescope but was even ignorant of the effects of each lens taken separately; and M. de la Hire, in the *Proceedings of the Academy of Paris,* in 1717, proved that Porta in the specious presentation of his *Natural Magic,* spoke only of a simple eyeglass, in which he had so combined a convex lens with a concave that they aided the sight of *those with blurred vision.* M. de Montuclas, always an excellent judge and apologist of Italian inventions, is of the same opinion, and says in his *History of Mathematics,* that before Galileo's time the telescope was unknown; Galileo endeavored always to perfect it, so that he invented one through which one could look with both eyes; he sent it in 1618 to the Archduke of Austria, Leopold. It is astonishing that Rheita, in a book printed in 1645, had wished to appear the inventor of it.

Use should be esteemed more than invention. The telescope in Holland was, like the magnet in China, an object of simple curiosity. Galileo in the same year 1609, watching the moon with the telescope, observed that the progressions of light after the new moon were irregular, sudden traces of light being emitted successively from the part still obscured. Not being enslaved to the prejudices of the old schools, he immediately understood that the moon was similar to our globe, and like it covered with valleys and mountains even higher than ours. Galileo, in his first dialogue on the system of the world, explained very well the resemblance between the two planets: it was (this resemblance) carried even further by other authors who discovered around the moon various indications of an atmosphere more rarified and variable than ours, and wished to explain in that way the circle which surrounds the moon at the times of the eclipses of the sun, and the variations which MM. de Mairan, Cassini, de la Hire, Maraldi, Kirk and de l'Isle have observed several times in the planets and the fixed stars next to the lunar disk; and Galileo, after the discovery of the telescope, still continued his lunar observation; for a few years before he lost his sight, as Viviani tells it, he discovered the libration of the lunar body by the observations which he made on the *Grimaldi* spot itself and on the *Mare Crisium* with which Grimaldi, Hevelius and Bouillaud were later so much concerned. The observation is described in the dialogue that we have cited, where it still seems that in number fifty-nine there was anticipated Newton's conjecture on the cause of the moon's always turning the same side toward the earth. One reads there that it is *evident that the moon, as if attracted by a magnetic force, always turns the same side toward the terrestrial globe and never changes.*

All the heavens seemed to offer Galileo new phenomena;

the Milky Way seemed to him made up of an innumerable quantity of very small stars: he counted more than forty of them in the single group of the Pleiades, and more than five hundred in the constellation Orion; the nebula of Orion alone seemed to him to be composed of twenty-two very small and closely spaced stars; that of Cancer, around forty: he also saw the four satellites of Jupiter, discovered the spots of the sun, the phases of Venus and Mars: he observed certain aspects of Saturn, which were later considered more at length by Huygens, who explained them by the hypothesis of a ring. Galileo carried his observations of Jupiter to the highest degree of refinement. After three years of work, he introduced the theory of the satellites, and at the beginning of 1613 he dared to predict all their configurations during two consecutive months. He then conceived of using them for the problem of longitudes; and in 1636, through the medium of Hugo Grotius, he offered, in behalf of the states of Holland, to work on that exclusively. The states gladly accepted his offer, sent Galileo a chain of gold, and delegated four commissioners to confer with him. Martin Hortensius, one of them, traveled to Tuscany shortly before Galileo lost his sight. Galileo, after this misfortune, communicated his observations and his writings to Renieri, who was charged by the Grand Duke to extend the tables and the ephemerides of the satellites of Jupiter. Renieri compiled them accurately, and showed them to the Grand Duke and many others, when, just as he was about to publish them, he lost his life in a sudden illness. I do not know by what accident his papers and those he had received from Galileo were lost.

The phases of Venus proved what some ancient astronomers had only supposed, that Venus did not move around the earth, but around the sun. Copernicus accepted this hypothesis, and even added that it was necessary that the phases of Venus resemble those of the moon. Galileo's telescope showed the resemblance of the phases of Venus, and some inequalities of Mars; phenomena which clearly proved the movement of Venus and Mars around the sun, and from which one can believe that the other principal planets likewise move around the sun. As M. de Montuclas has very aptly commented, what would have been the joy of Copernicus if he had been able to adduce similar proofs in his favor! Galileo contributed much through his *Dialogue on the Great World Systems* to the triumph which the illustrious Prussian's system has since won, and which was so baneful to our Italian. In the second dialogue the terrestrial phenomena are so well explained; and in the third, all the celestial; the simplicity of Copernicus' hypothesis is so well upheld, and the difficulties of other hypotheses of Ptolemy and Tycho Brahe explained so clearly, that one begins to understand the movement of the earth through his dialogues with as much certitude as one could have in physical matters, even before Bradley, in England, had discovered the aberration of light, verified in Italy by Eustachio Manfredi who will always live in the history and annals of astronomy.

Galileo, before leaving Padua, had discovered the sunspots; and being in Rome in the month of April 1611, he had shown them to several distinguished persons who verified the fact. The first observations of Scheiner were made six months later: he announced them later in 1612, under the title *Apelles post tabulam,* with three letters addressed to Velser. Galileo answered immediately, and assured for himself the honor of having first discovered these spots. He also made plain that the fictitious Apelle had offered a theory the very opposite of his, in claiming that the spots move from East to West, and that they decline towards the South; while really they move from West to East, and decline toward the North. Perhaps Apelle, attached to the old opinion of the incorruptibility of the heavens, thought that these spots were planets. As for Galileo, who was a man above all prejudice, he says in his letters to Velser, that these spots consisted of matter very close to the sun's surface, which gathered and dissipated, and reproduced in the manner of vapors of our atmosphere;

and he judged by the movement of these spots, that the sun made a complete revolution in about the time of a lunar month. M. de Montuclas has accorded to Galileo the honor of having, although the first, spoken more judiciously than the others on these spots.

It was in the year 1612, that Galileo began to publish his discoveries concerning sunspots in his work on bodies that float or move on a fluid. He re-established by this discourse the hydrostatic doctrine of Archimedes, and demonstrated that the sinking of solids in a fluid, or their floating, does not depend at all on the shape of the solids, but on their specific gravity. In the work entitled *The Assayer*, which the Count Algarotti recognized as the best polemic of which Italy can boast: in this work, I say, it is formally established as a maxim, that the observable qualities, such as color and taste, do not reside really in bodies, but in ourselves; a maxim that one should attribute more to the ancient philosophers than to Descartes. Thus Galileo fixed the principles of hydraulics and physics: he was the first to create mechanics. In the year 1602, he wrote to the Marquis Del Monte that he had observed that the oscillations of movable bodies attached to wires of different lengths are made in times proportional to the square roots of their lengths. He announced in a letter written from Padua, in 1604, the theorem that the distances that heavy bodies cover while falling are as the squares of the times, and that nevertheless the distances that they cover in equal times are as 1, 3, 5, 7, etc. The first edition of his dialogues on mechanics appeared in 1638, the same year as Baliani's treatise on movement; but the writings and the discoveries of Galileo in mechanics had spread far and wide well before this time; and it is not likely that Descartes, and even less Baliani, would have discovered much in mechanics without having read Galileo.

Among the principal discoveries found in his dialogue on mechanics, I accord first place to the principle of the composition and resolution of motion, that Galileo expressly taught in the second theorem on the movement of projectiles, and in the note to the second theorem on accelerated motion. I accord second place to the laws of uniform motion and of accelerated motion, from which result the well-known formulas commonly called *the formulas of Galileo:* 1. that the force multiplied by the component of the time is equal to the component of the speed; 2. that the force multiplied by the component of the distance is equal to the component of the speed multiplied by the total speed. Galileo considered these two formulas in the case of constant force, and Newton then extended them generally to all hypotheses of variable force. But everything that has been done since in mechanics depends entirely on these two formulas and on the principle of the composition and resolution of motion. The treatment of motion on inclined planes and in chords of circular arcs is full of geometric elegance; and it will always be astonishing that one man alone reached this point without the aid of algebra. The problems in which one seeks the inclination of planes by which a body can travel the fastest, either from a given point to a horizontal line of given position, or from a horizontal line to a given point; these problems are of the subtlest order.

Galileo in his fourth dialogue has marvelously treated ballistics, which were totally ignored before him; for Cardan and Tortaglia suspected only that hurled projectiles travel in a line composed of a straight line and a circular arc. Galileo, with the principle of the composition of motion, demonstrated not only that projectiles describe a parabola, but even taught all that applies to the extent of the thrust, range, height and direction; for with two of these quantities, one can always derive the two others. Finally in the second dialogue he revealed further the principles of the whole doctrine of the resistance of solids, which was later very fully developed by Viviani and Grandi.

Galileo, in his first and third dialogues, while treating the cylinder carved out of a hemisphere, and the distances covered in accelerated motion, left us traces of the method of indivisibles, in considering solids as composed of an infinitude of planes, and the planes as an infinitude of lines. But the truth obliges us here to observe: 1. that Kepler had, in his *Stereometry*, already introduced infinity into mathematics, and furnished the idea of indivisibles; 2. that Cavalieri used with much precaution these same metaphysical phrases, as he showed in the preface to book seven of his *Geometry*, and as Maclaurin observed; 3. that, although Galileo planned to compose a *Geometric Treatise* on indivisibles, he had no part in the great work of Cavalieri. One could add to the many proofs that one has of this, that of a letter which Cavalieri wrote to Galileo on March 21, 1616, which is an incontestable proof that the first had ended his work before the other had so much as begun his. *In regard to the work on indivisibles*, he says, *I would be pleased if you worked on it as soon as you can, so that I could expedite mine, which I will rework while waiting*, etc. Cavalieri published his work three years later, and it was the principal basis of the differential and integral calculus.

But to return to the dialogues, in the first edition, and in the third of his dialogues, Galileo gave as an axiom, that a moving body passing from a given point on any inclined plane to a given horizontal line, always arrives there with the same speed. Viviani was the first to make it evident that this principle has need of some demonstration; and Galileo, although blind, found it right away, and informed Viviani in the way that we see in the other editions of his dialogues. Galileo, in his *Discourse on the Bisenzio River*, applied this proposition to the case of two currents; and explained in another theorem, that the speeds are the same in two canals of different lengths and of different curvatures, when they are the same depth, that is, when they remain fixed in the same limits. In the particular case of rivers, other factors enter into consideration; but the proportion generally taken is quite true, and the application that Galileo made, the first of geometry to the knowledge of water currents, brought him much honor.

Varignon has uncovered an error which is in the seventeenth theorem of the third dialogue, where Galileo supposes that a body passing from one plane to another of a different inclination retains entirely the velocity corresponding to the first fall; but Grandi, in his notes to the same dialogue, says that Galileo's passage should not be taken in an absolute sense, but as a simple hypothesis from which it is necessary to depart in order to arrive eventually at the fall of bodies in circular arcs.

It is very true that in circular arcs, as in all the curved lines, there is no point of observable alteration in relation to the different inclinations of the small arcs of which the curved line is composed, as Varignon, Grandi, and many others have demonstrated. One can not find a more marvelous theorem than that by which Galileo blazed a trail, his hypothesis that a body descends faster along a circular arc than along a chord. Jean Bernoulli has interpreted the theorem too generally, as if Galileo had believed that the descent is faster along a circular arc than along any other curved line comprised between two given points; later Bernoulli proved that the curve of the fastest descent is a cycloid, and not a circular arc. But the note to the twenty-second theorem suffices to show that Galileo wished to say only what is indisputable: *Therefore, as we approach closer to the circumference with inscribed polygons, the motion between two given end points is completed more quickly.*

Nevertheless, the belief has been generally imputed to Galileo that the curved parabolic line, in which a projected body moves is the same as that to which a chain suspended by its ends, and is called a catenary, conforms itself; and it is significant that Krafft, in his last days, should have attempted to defend this, in book five of the *New Commentaries of Petersburg*, citing the passage which follows the fourteenth proposition of the fourth dialogue, which says only that the two curves do not differ much from each other. "The extended cord, more or less stretched, bends in a line which rather approximates some parabolics; and the resemblance is such, that if you mark a parabolic line on a plane surface elevated on the horizon, and hold it upside down; that is, the summit to the bottom, and with the base parallel to the horizon, holding suspended a small chain held at the extremities of the base of the marked parabola, you will see the said chain, while more or less being released, curve and adapt itself to the same parabola; and this adaptation is more precise as the marked parabola is less curved; that is, more stretched; so that in the parabolas described with the elevation of 45°, the chain follows along the parabola almost to a hair.

Galileo turned a little later to another proposition. Given a horizontal cord turning on two pivots, and taken as weightless, that is pulled straight by two very heavy weights attached to the ends; if another weight, regardless of how small, be attached to the middle, it will bend there, and consequently will no longer be straight. Viviani, writing to the prelate Ricci, raised some doubts about Galileo's demonstration, based especially on the fact that the movement of the two weights in rising when the cord bends is not equal. This objection, although supported by some famous men, does not seem to be applicable to Galileo's case, in which, assuming weights that are infinitely large, in consideration of the little body attached to the middle of the cord, their movement could be only very slight and consequently uniform. It is true that the state of equilibrium is not precisely that assumed by Galileo in his demonstration, as Viviani suspected, and as Simpson pointed out in the thirty-eighth problem of the application of algebra to geometry. But Galileo's demonstration can be adapted equally to the true state of equilibrium, and the principal proposition remains quite true. To these mechanical difficulties, there are joined several others, physical and astronomical, which can be reduced principally to three: 1. that Galileo attributed the elevation of water in pumps to abhorrence of the vacuum; 2. that he wished to explain the rise and fall of the sea by the combination of the daily and annual movement of the earth; 3. that he did not believe that comets were planets which turned around the sun. As to the first objection, Galileo, in the first dialogue, described simply this phenomenon that water rises only thirty-two feet in pumps, and simply inferred from this that the force necessary to overcome the vacuum, equals a column of water thirty-two feet in height; and against that there is nothing to say, although Galileo had added other conjectures which are not equally sound. Galileo, in addition, proposed a machine to measure how much greater the force of cohesion is than that which one requires to create the vacuum, and later he provided two different methods for measuring the very weight of air; and although in his experiments he did not deduce any proportion between the weight of air and water, which is of one to four hundred, his experiments should be regarded nevertheless as the foundation and the principle of all that has since been done on the subject.

The hypothesis given in the fourth dialogue on the system of the world, to explain the rise and fall of tides, is very ingenious, and it is the first by which scientists have attempted to explain physically this singular phenomenon; and although the hypothesis is not true, Descartes, who has written since Galileo, has not given a better one. In regard to the comets, Galileo objected against his adversary, that he had not yet proved that comets were solid and inalterable bodies, and that parallax can serve to measure the distance of bodies, but cannot be applied to simple optical appearances, among which comets can be counted. Cassini also has upheld, in a book printed in 1653, and dedicated to the Duke de Modene, that comets are a mass of exhalations from the earth and planets. It is a short time later, as M. de Fontenelle remarks, after Cassini had found the irregularities of the movement of comets to be purely apparent, and the comets themselves, like the planets, to be subject to calculation, that astronomers began to believe, on good grounds, that comets are solid bodies, and that, even like the other planets, they turn around the sun.

M. de Fontenelle, in his eulogy of Viviani, regards Galileo as a rare genius, whose name will always stand foremost with the discoveries of greatest importance on which science is founded. Descartes, so inferior to Galileo, blamed him precisely for what was most praiseworthy; that is, for being satisfied with facts and demonstrations, and for not seeking to arrive at first causes. Newton, whose genius was more than human, has perhaps fallen into more errors than Galileo. We must admire in Galileo a philosopher, a geometer, a mechanician and an astronomer competent in the practical no less than in the theoretic sphere; the very one who has dissipated the errors of former thought; the soundest and most elegant writer that Italy has produced; the master of Torricelli, Castelli, Aggiunti, Viviani, Borelli, Paolo and Candido del Buono. The last four are the ones who founded the Academy of Cimento, of which the essays, worthy of Newton's century, seemed written by the genius of Galileo, as it appears in the preface of the *Proceedings of the Academy of Dijon,* cited at the beginning of this essay.

from

Galileo*

THE RIGHTS OF SCIENCE
AND
THE METHOD OF THE PHYSICAL SCIENCES

by

THOMAS HENRI MARTIN

[Martin's work is generally acknowledged by serious scholars to have been the first competent critical examination of the nature of Galileo's achievement both as a scientist and as a defender of the rights of science. The book from which the following reading is drawn offers, in an appendix, an extensive bibliography with ample annotations, assessing the contribution of previous scholars to the critical examination of the subject. Martin's own account of Galileo's involvement with the Church on the rights of science in relation to religion—an account based on thorough examination of many documents which he was the first to examine—is referred to by White, in his *A History of the Warfare of Science with Theology in Christendom*, as a "fair summing up of the case."]

INTRODUCTION

Because of the studies to which he devoted all his life, because of his discoveries and his writings, Galileo holds an eminent place in the history of the physical sciences, especially in mechanics, optics and astronomy. Not only did

---

* Translated by James Brophy, from Thomas Henri Martin, *Galilée*, Paris, 1868. Copyright © 1961 by James Brophy.

he enable these sciences to make decisive advances, but he prepared for their future progress by inventing many of their most necessary instruments, and, most of all, by supplying them with their true method. He did not formulate this genuinely philosophic method in a body of precepts; but he gave excellent examples of it, explained and justified by him in his writings and thus took the best way of establishing and propagating it. By this method, more complete, simple and much more efficacious than that of Francis Bacon's, and by the philosophic principles that it assumes and which Galileo well understood, the Florentine scientist should finally occupy a higher place than the Chancellor of England in the history of modern thought. This position is due to him also, because with his intellectual superiority, his zeal for science and his sincere respect for the Faith, with his vigorous argumentation and his mordant irony against prevailing errors, and with his apparent submission to abusive prohibition, he defended to the best of his ability, against an intolerance excited and deceived by the claims of pseudo-scientists, unworthy rivals and stubborn enemies, the legitimate liberty of the human spirit in science, and because, accused and condemned for a true doctrine, forced to abjure it and confined thereafter for the rest of his life, he suffered greatly for this noble cause which he saw oppressed for a time, but which he knew would be victorious in the end.

After more than two centuries, the calm necessary for impartiality finally begins to settle around the memory of Galileo. . . . Yet despite his submission some have not yet been able to pardon him, and wish absolutely to find him guilty of temerity, heresy, and imaginary malice; others would go so far as to transform the judicial error of an ecclesiastical tribunal into a dogmatic error of the Church, the humble and submissive attitude of Galileo before his judges into an heroic resistance, and to add to the representation of the trial and its aftermath some revolting touches: for the one, the historic truth about Galileo is unpleasant; for the other, it is insufficient. Galileo, also, is Italian: Italy has its admirers and its passionate enemies. Descartes was a great philosopher and a great mathematician; Bacon, a felicitous writer especially in the *Essays*, was nothing in mathematics and mediocre in philosophy; neither was a great physicist. Certain temperaments would have it that one or the other, and not Galileo, had established the true method of the physical sciences, while others, at the same time that they lengthen, at the expense of truth, the list of the discoveries that this method enabled the Italian thinker to make, ignore or conceal the errors that he committed in applying it—errors which, produced by certain too exclusive preoccupations, are a tribute paid by this great mind to human weakness.

Galileo was condemned by the Roman Inquisition: that is why certain Catholics, more ardent than enlightened, have been led to find him guilty; they have not forgiven him the injustice he suffered. Galileo, in spite of his faults, was a good Christian all his life, at least in his own belief, and this is the reason why, according to Trouessart, he received but little sympathy from our French Encyclopedists and their followers. These men, so indulgent to the voluntary and sacrilegious hypocrisy of Voltaire, found that Galileo had put himself in the wrong theologizing in 1615, and that he had been weak confronting the intolerance of 1633. To please them he would have had to attack Christianity, or at least let himself be burned by the Inquisition as a heretic.

In the formula of abjuration, after having admitted that his dialogue is ultimately favorable to the false doctrine of the movement of the earth and of the immobility of the sun, and that in publishing it he had violated the absolute injunction, made against him in 1616, to no longer defend or teach in any manner this doctrine declared contrary to holy scripture, he *affirms and swears, his hand on the holy Bible, that with a sincere heart and a strong faith, he recants, de-*

nounces, and detests the above errors and heresies of the immobility of the sun and the movement of the earth, by which he had been justly condemned as *vehemently suspect of heresy.* Not only does he promise to abstain henceforth from all heretical doctrine, but he *swears that, if he would come to know any heretic or any one suspected of heresy, he would denounce him to the Holy Office if not to the Inquisition and the ordinary of the area.*

While the unfortunate Galileo, to escape a punishment that he had not merited, signed and read this work of the Inquisition, certainly his oppressed reason protested interiorly in favor of the truth that he was forced to deny again; his oppressed conscience protested against the odious promise to renounce to his oppressors his disciples, reputedly *heretics* or *suspect of heresy,* because they would continue to see clearly, as he, a question of astronomy. If, instead of consenting to these humiliating lies, Galileo had braved, without being a heretic, the punishment pronounced then by the inquisition against relapsed and impenitent heretics, that is to say, the stake, we would have to admire him perhaps, at the same time deploring the immense evil that would have resulted from his intrepid resolution. Since he did not have this baleful courage, we must excuse him, and not hold him responsible for the blindness of the judges who forced him to speak against his conscience and against the truth: we must try to understand, as we have, the motives of universal interest which, together with personal interest, may have led him to this act of apparent submission.

But, in going through with this exceptionally blameworthy act, Galileo would not have been able, without absurdity, to give himself the miserable satisfaction of contradicting it at the risk of destroying himself. No one knows who invented the tale repeated again and again down to our own time, according to which, as soon as he had abjured on his knees the movement of the earth, Galileo had gotten up saying "*E pur si muove.*" The only possible witnesses to those supposed words of Galileo would have been his judges. "Nevertheless the earth does move": certainly Galileo thought it, and the inquisitors, who forced him to lie, well knew that he thought it, but he would not have said it in an audible tone with impunity; if he had said it in that way, his detention would not have ended two days later. There is no proof that, in speech or in writing, he ever dared by overtly professing the system of the earth's movement to contradict expressly his abjuration.

Galileo, after his condemnation and his forced recantation, always believed completely in the truth and ultimate triumph of his system, and in its compatibility with the Catholic faith: he thought that in defending it, in 1632 as in 1615, he had not violated his duties as a Catholic. In several of his letters, written to correspondents in whom he had complete confidence, for example in his letter of July 25, 1634 to Diodati, in that of February 21, 1631 to Peiresc, and in a letter of 1637 to the king of Poland, he energetically expressed, and with a true eloquence, his profound conviction of his innocence from a religious point of view. After having affectionately thanked Peiresc for his generous intervention to obtain the reduction of his punishment, he adds, "I have said . . . that I do not hope for any leniency; and that, because I have committed no wrong; I could hope for and obtain mercy and pardon, if I had transgressed; for it is upon transgressions that the prince can exercise mercy and indulgence, while towards one condemned in innocence it is necessary to maintain rigor in order to show that one is certain to have acted according to law. But (believe me for your consolation) all this afflicts me less than one could believe, for I am never without two consolations: one is that, in the reading of all my works, no one could ever find the least shadow of a thing that deviates from piety and reverence toward the Holy Church; the other is my own conscience, which is fully known only by me on earth and by God in heaven, and which knows that, in the cause for which I suf-

fer, many others would have been able to argue and speak
more wisely, but no one, even among the holy Fathers, could
have done so more piously or with more zeal for the Holy
Church, or, finally, with an intention purer than mine." Such
were the feelings of the condemned Galileo, sentiments the
sincerity of which could not be doubted; for they accord
with all his conduct, with all his words, with all his letters
and all his writings. Certainly, after a sentence like that to
which he was subjected, after an abjuration like that which
he was forced to make, and in the penal situation where
we see him kept, it is to Galileo's credit that he remained
devoted to the Catholic religion.

In all his works his style is habitually clear, simple, natural,
at times lively and piquant, sometimes grave and elevated,
but often a little verbose. In his two principal works, he has
used the dialogue form. Necessarily a little prolix, this form
is useful to present questions in all their aspects, and it is
favorable to polemic, which Galileo needed to make his ideas
prevail against the opposition of the obstinately prejudiced
and of his hateful rivals. In his *Dialogues Concerning Two
New Sciences,* the tone is calmer and the form more didactic:
it is quite so even in some long mathematical demonstrations
introduced in the form of citation. But it is especially in his
Astronomical Dialogue, a work basically less perfect, that
Galileo has utilized all his resources for this *genre* of writing:
he used there with success, and as he said himself, in imitation of Plato, what Socrates called the method of dialectical
midwifery; that is, a method which consists of leading the
adversary, by a series of adroitly presented questions and
of easily obtained responses, to admit what he thought he
did not know or even what he held to the contrary, or even
to admit, in perceiving his concessions, that he did not
know what he said and that he did not even understand
himself. Furthermore, the dialogue form served Galileo here
to have another person say that which misinformed intolerance would not have permitted him to say in his own name.

Not only in his two great works of dialogue, but also in his *Defense Against Capra*, in his *Assayer*, in his shorter works, his notes and his polemical letters, Galileo showed a remarkable talent for dialectic with much verve and irony. Especially in his *Dialogue on the Great World Systems*, he, as a genuine philosopher, has given to the peripatetics some just and severe lessons of logic applied to the physical sciences. In his two apologetic letters to Father Castelli and to the grand duchess Christine, he repelled, with the same superiority of reason and of knowledge and with the same literary skill, the theological attacks directed in the name of the Bible and Aristotle against the new system of the world.

However, in criticizing the inadequate physics of the Peripatetics, he recognized in Aristotle's works on the logic and utility of deductive reasoning a merit that Bacon had refused them; but Galileo remarks that one could be very strong in logical theory, and weak in some applications, where the logic would give an error to the method and to the opinions of the author: he cites as examples of this some of Aristotle's reasoning in physics.

Galileo repells with as much energy as reason the pretentions of those who wish to resolve physical questions by the principle of authority. In 1612, the peripatetic Lagalla having opposed the universal consent of men to the doctrine of the double movement of the earth, Galileo, in his note viii, answers him with the adage: "*Stultorum infinitus est humerus*"; and here is his commentary on this proverb. "The whole of philosophy is known only by one being, who is God; as to those who know something of it, their number is as small as the knowledge they have of it; but the greatest, one might say infinite, number falls to the ignorant." Father Grassi, in his *Astronomical Balance*, a violent pamphlet against Galileo, used the same argument as Lagalla. Galileo, in the *Assayer*, gave him, with some changes, the same answer. Father Grassi had cited numerous texts of ancient

poets and writers to prove that a lead ball launched by a
sling is heated to the melting point in the air, and that the
Babylonians cooked eggs by twirling them in their slings;
then he used the alleged authority of several ancient philosophers to show that such must be the result of rapid
movement. Galileo, in his *Assayer,* rejected the announced
facts, while remarking that an error repeated a hundred
times does not become a truth, and that no authority is
needed for an assertion that one can refute every day by
experience. So much for the facts. As for the theory, Galileo
says that the authority of a single competent man, who gives
good reasons, is worth more than the unanimous consent of
those who understand nothing about it. In effect, he says,
"If the action of discourse on a difficult problem were like
the work of carrying baggage, work in which many horses
could carry, for example, more sacks of grain than a single
horse, I would agree with your opinion that several search-
ers would be better than one; but the work of discovery is
comparable to that of running rather than carrying, and a
single Arabian steed will run faster than a hundred Frisian
horses." Galileo also knew that the authority of a man, even
of a great mind, is worth nothing against definite proof.
It is necessary to see with what derisive and eloquent verve,
at the beginning of his *Astronomical Dialogue,* he battles
these obstinately servile peripatetics, who, for example, in
favor of their dogma on the immutability of the heavens and
the stars, did not hesitate to oppose some texts of Aristotle to
incontestable observations, which, as Galileo says, Aristotle
would have accepted, if he had known them: "It was his
partisans," he says, "who claimed authority for him, and not
Aristotle himself; and because it is easier to hide behind
another's shield than to appear unprotected, they are afraid
and dare not take a single step, and, rather than make any
alteration in Aristotle's heaven, they impertinently wish to
deny what they see in the sky of nature."

But, in spite of this independence of spirit, Galileo is very far from having professed, as Descartes and Malebranche, a sovereign disdain for the study of the doctrines of the earlier great philosophers. On the contrary, he had extensively used such study, and he declared to have profited thereby, because he had known how to guard his freedom of judgment. What he blamed was the abuse of investigating only books instead of observing and reasoning for one's self. He kept himself carefully between a proud isolation from and servile subjection to another's thought. It was the latter of these two excesses that prevailed around him; it is this excess that he fought in all his polemics against the peripatetics, for example in his Note I on the *Discourse of Lagalla:* "Between philosophizing and studying philosophy, there is," he says, "the same difference as between drawing from nature and copying the drawings of others." Then he states that the study of philosophical works is very useful in exciting and directing minds. But he remarks that an artist who always limits himself to copying without ever attempting to draw from nature would never become either a good painter or a good critic of painting. "In the same way," he says, "while always working with the writings of others, and in so dissipating one's efforts, without ever raising one's eyes to the works of nature to seek and discover there the truths already found and to follow the evidence of some of these infinitely numerous truths which remain to be discovered, one will never be a scientist, but an amateur versed in knowledge about others' writings on science."

All these views of Galileo are prior to the publication of Descartes' *Discours sur la méthode.*

In the philosophy of science, Galileo made the physical sciences the principal object of his studies. But philosophy itself is familiar to him. He gives proof of this especially in his *Dialogue on the Great World Systems.* One finds there, for example, on the first day, an excellent discussion on the position to hold between the suspect and dangerous humility of scepticism, which by denying all absolute authority to reason makes all science impossible, and the feeble pride of extreme rationalism, which claims to know and understand everything, and which puts itself in the place of God by an illusion fertile in deplorable errors. In this matter, Galileo expresses with justice the infinite difference between the divine intelligence and ours, in regard to the range and way of knowing: he shows us man painfully acquiring by successive effort of observation and reason a few perfectly ascertained but limited assumptions, and others more or less probable, which joined with the first embrace only a very small part of the universal truth, while God knows all with complete certitude by a single eternal intuition without succession of thoughts. During the third day, on the grandeur of God's works, and our inability to understand them, Galileo finds and gives to one of his disciples some eloquent and true expressions which Pascal or Bossuet would not have rejected. We have already indicated a passage of the fourth day, in which he states that beyond all secondary causes, it is necessary to discover a first cause, of which an action all-powerful, infinitely wise, and which he declares essentially miraculous—the creating action—can alone explain the origin of all things.

We have said that Galileo gives great weight to the consideration of final causes: in this same dialogue, one sees that he even possessed the theory of it, and that he understands the application of it without exaggerating it. He especially does not wish that utility for man on earth be considered the unique end of all things in the world, nor that a final cause, however real it may be, be considered the only end of the natural power to which it relates; for, according to Galileo, divine Providence, general and special at the same time, functions entirely together and also entirely apart, in the same way, he says, as the sun, which fills all our planetary system with light and heat, matures the little grape as effectively as if the end of its action were exclusively the growth of that fruit.

As is evident, Galileo, apart from mechanics, physics, and astronomy, easily knew how to find the important ideas in pure philosophy and express them in a language as true and simple as they are. The range and height of his intellect were graced with the exactitude of his method and the compelling rectitude with which he applied it. A religious philosopher at the same time as a mathematician and physicist, he was in the great tradition of Pascal, Newton, Ampère and Cauchy.

We have seen how Galileo has played the principal role in the creation of this excellent method, which he followed all his life with as much success as constancy, and which, perfected and developed after him, will remain that of the physical sciences. We have seen that the eminent astronomer Kepler owed his most brilliant discoveries in astronomy and in optics to this method, but he followed it only after his contemporary Galileo, and too often he deviated from it in rash speculation. We have seen that Descartes, another contemporary, completely misunderstood the procedures and results of this method, and that in physics he led himself astray by refusing to follow it. As for Chancellor Bacon, we have seen that before the publication of his *Novum organum scientarum,* he knew the works in which Galileo, before this period, had professed and demonstrated the experimental method. It is not without cause that Bacon called himself the *trumpet* (*buccinator*) of this method. But, in proclaiming it, he altered it, mutilated it, and happily the true scientists have not followed the lead of this unfaithful herald.

Since Galileo, all the important physicists employ his method, the first operation of which is not only to observe and experiment but also and especially to measure and count, after having for convenience, by mathematical transformation, shifted the problems of physics into the research of measurable quantities. Next, aided by the principles of reason and mathematical process, this method leads to the understanding of the mathematically precise laws which govern phenomena, and permits the advance, when possible, from the measurement of effect to the knowledge of causes and their first laws. This is the great secret of true physicists, the secret of which Bacon, too removed from mathematics, understood only a part, and even that imperfectly, and of which the positivists, more mathematical, would desire to suppress only the last and essential point.

The national pride of the English would have excused their presenting Bacon as the renovator of the physical sciences; but no: our French Encyclopedists are those who especially have wished to assure this honor to the one whom they consider the father of their sensualist philosophy. This explains, especially in France, the forced vogue of Bacon as a philosophical physicist. But, in the eighteenth century, the Englishman Hume had announced in good faith the superiority of Galileo's rights to the title renovator of the sciences. Undoubtedly, there is some exaggeration and injustice in Joseph de Maistre's two volumes of accusation against Bacon's philosophy; but there is also much truth, particularly concerning the Lord Chancellor's erroneous physics and his absurd astronomy. M. Charles de Remusat inclines to give Galileo the principal role in inventing the *practice* of the true method of the physical sciences, and Bacon the credit for the invention of the *theory* of this method. But M. Cournot evaluates the just value of "this prolix enumeration of *instances* and *forms of induction,* to which Bacon attached as much and more importance than the Scholastics attached to the syllogistic forms, and of which, according to M. Cournot, "no one after him has made any use." In the same manner, while saying that "the general maxims of Bacon are sagacious and conducive to the stimulation of minds," the English scientist Whewell recognized that "his particular precepts slipped through his hands and are now of no use in practice." In saying *"now,"* Whewell says too little. Liebig, the German chemist, declares, as does Cournot, that physical scientists have *never* proceeded according to the rules of the *Novum organum scientarum.* The eminent English critic Macaulay, and after him Apelt, have clearly shown that the most false inductions can satisfy the greatly complicated and greatly insufficient laws of Baconian induction, and that a few examples of a legitimate and well-conducted induction are worth more than all this new scholasticism. But Galileo and Kepler had given excellent practical examples of induction, and Bacon did not know how to imitate or understand them: he knew only how to scorn them. Contrary to the opinion of Macaulay, and in accord with the scientists whom I have cited, I think that the *practical weakness* of Bacon's method only betrays the *theoretical error* of this method, which recommends itself for several excellent maxims and for some ingenious and true insights, but which is defective in its entirety, and which ignores what was most necessary to prescribe; that is, first, mathematical exactitude and the use of precision instruments in observations, and secondly, mathematical procedures in the inductions that one makes to arrive at precise laws and causes. In 1816, the French physicist Biot spoke energetically on the established inutility of Bacon's method and on the perpetual use of Galileo's. The illustrious English physicist, Sir David Brewster, tells us: "If Bacon had never existed, he who studies nature would have found in Galileo's works and writings, not only the much vaunted principles of inductive philosophy, but also their practical application in the highest efforts of invention and discovery." Such are, on Galileo and Bacon, the judgments of the most competent physicists, who have examined the question without prejudice of philosophic school or national pride. One of them, Trouessart, after having shown that Galileo's method is not Bacon's, summarizes his judgment in these words: "In science, we are all disciples of Galileo."

As for metaphysicians, they can only profit by studying this method, to which they have given so little attention up to now; for it accords perfectly with their principles, although it is not made for their science, to which Galileo never pretended to apply it. He knew that physical things can be measured, but moral things can not. Bacon, on the contrary, although himself spiritual and even a bit metaphysical, in advocating too exclusively the method of observation, in minimizing the role of intuitive reason, metaphysics, mathematics and deductive reason, prepared for the similarly too exclusive application which was made of the experimental method in philosophy, first by the timid and moderate sensualism of Locke, then by the whole sensualist and materialist school of the eighteenth century and by the continuators of this school in the nineteenth. This school has accomplished all that was in its power to falsify, vilify and almost annihilate philosophy, when under the pretext of accepting only phenomena observable by the senses, it recognized man no longer as a thinking, free, responsible and immortal being, but only as a variable sum of physical phenomena and of sensations more or less transformed, without unity, without actual substance, without persistent identity, without relation to causality in the present, past and future. Furthermore, in diminishing the legitimate field of the human spirit, in shattering the ideas of cause and substance, in obscuring the idea of active force and finality in nature, this same school, so deadly to philosophy, would have done almost as much harm to the physical sciences, if it had not been happily illogical. In effect, its theories, sensual and ideal at the same time, should have been able to lead it to doubt the objective reality of bodies, and consequently to repudiate Galileo's method, in order blindly to record and classify sensations. However, the majority of physicists of this school, despite their erroneous philosophy, have kept approximately intact in the practice of their science the method of Galileo, this method which assumes what they deny, and which recommends in fact what in theory they pretend to proscribe. Thanks to this method, which they have rather conserved by tradition despite their doctrine, they have continued to believe for practical purposes in final causes, while repudiating them; they have continued to rely on the principle of efficient causality and to advance in the knowledge of forces by the measurement of their effects, in the knowledge of beings by the observations of the manifestations of forces which act in them; they have continued to proceed in this manner even while arguing the necessity to abstain from the search for cause and effect, and to be concerned only with phenomena, their classification and their order of succession.

Thus, Galileo's method, even as his doctrine on the system of the world, has triumphed over all opposition raised against it. Open to all progress, it remains dominant in the field of science for the present and for the future, which will continue to add new perfections to it while retracting nothing. The use of this physico-mathematical method had begun in antiquity for some aspects of the mechanics of solids, of hydraulics, acoustics, optics, and astronomy. In spite of the strange ideas to which we have referred, Kepler made an excellent application of this method to determine exactly the movements of Mars, and then to determine the elliptical revolutions of all the planets and the geometric laws of the revolutions. Then, by the same method, Newton reached the mechanical principle of these laws; he thus found what positivism would have forbidden him to search for; that is, the *cause* of planetary movements; instead of searching, as Descartes would have directed him, for the *essence* of this cause, he found the mathematical *law* of the action of this, until then unknown, *force*, and because he knew this force and its mode of action, without having need to decide if it is irreducible or even if it is only a manifestation of a more general force, he was able to go further than Kepler's laws in confirming them; he was able to rectify the third law, which is exactly true only to the extent that the mass of the planet is a negligible quantity in relation to the mass of the sun; he was able, further, to establish the fundamentals of the theory of the perturbations caused by the mutual attractions of the universe's various bodies: deviations henceforth calculable, which fortunately the imperfection of instruments had prevented Kepler from noticing; for perhaps they would have made him doubt his laws, as Newton doubted his in 1666, but, in his case, because of the inaccuracy of hypotheses on which he operated. Thus it is that by following Galileo's method to the end; that is, in rising from secondary laws, the first result of experimental induction aided by measurement and the calculus, to the first laws of motivating forces, Newton arrived at celestial mechanics, an immense work, which continued after him and which will continue always.

It is Galileo who first solidly established this method, who extended it, who generalized and regularized the use of it, and who first showed the necessity of applying it to all the physical sciences. He employed it with great success himself in astronomy, for example in the study of sun spots, the mountains of the moon, the phases of Venus, the variations of the visible diameters of Venus and Mars, in his long and patient efforts for the determination (too difficult then) of the movements of the satellites of Jupiter which he had discovered, and in his discussions on the system of the world; he successfully used this method in all of the mechanics of solid bodies and in aspects of hydraulics, dioptics, magnetism, acoustics, etc. Without doubt, he made some errors, easy to discover even in his best works, for example in what concerns the cause of tides, the origin of comets, and the supposed resistance of the vacuum; but these mistakes, easily corrected by the use of his method, are considerably less numerous and considerably less serious than those which have been imputed to him by M. Arago, on the strength of citations and interpretations, which M. Alberti has justly demonstrated to be false.

from

A History of the Warfare of Science with Theology in Christendom*

by

ANDREW DICKSON WHITE

[White's work hardly needs an introduction. It was the culmination of many long years dedicated by the author to the defense of free inquiry against religious dogmatism. *The Columbia Encyclopedia* notes that, while teaching at the University of Michigan, he envisioned the establishment of a "university free to pursue truth without reference to dogma," and that, as first president of Cornell (1867-85) he was one of the first to break the dogmatic rigidity of the traditional college curriculum by introducing the use of free electives.]

All branches of the Protestant Church—Lutheran, Calvinist, Anglican—vied with each other in denouncing the Copernican doctrine as contrary to Scripture; and, at a later period, the Puritans showed the same tendency.

Said Martin Luther: "People gave ear to an upstart astrologer who strove to show that the earth revolves, not the heavens or the firmament, the sun and the moon. Whoever wishes to appear clever must devise some new system, which of all systems is of course the very best. This fool wishes to reverse the entire science of astronomy; but sacred Scripture tells us that Joshua commanded the sun to stand still, and not the earth." Melanchthon, mild as he was, was not

* Andrew Dickson White, A *History of the Warfare of Science with Theology in Christendom* (New York: D. Appleton and Company, 1896).

behind Luther in condemning Copernicus. In his treatise on the *Elements of Physics*, published six years after Copernicus' death, he says: "The eyes are witnesses that the heavens revolve in the space of twenty-four hours. But certain men, either from the love of novelty, or to make a display of ingenuity, have concluded that the earth moves; and they maintain that neither the eighth sphere nor the sun revolves. . . . Now, it is a want of honesty and decency to assert such notions publicly, and the example is pernicious. It is the part of a good mind to accept the truth as revealed by God and to acquiesce in it." Melanchthon then cites the passages in the Psalms and Ecclesiastes, which he declares assert positively and clearly that the earth stands fast and that the sun moves around it, and adds eight other proofs of his proposition that "the earth can be nowhere if not in the center of the universe." So earnest does this mildest of the reformers become, that he suggests severe measures to restrain such impious teachings as those of Copernicus.

While Lutheranism was thus condemning the theory of the earth's movement, other branches of the Protestant Church did not remain behind. Calvin took the lead, in his *Commentary on Genesis*, by condemning all who asserted that the earth is not at the center of the universe. He clinched the matter by the usual reference to the first verse of the ninety-third Psalm, and asked, "Who will venture to place the authority of Copernicus above that of the Holy Spirit?" Turretin, Calvin's famous successor, even after Kepler and Newton had virtually completed the theory of Copernicus and Galileo, put forth his compendium of theology, in which he proved, from a multitude of scriptural texts, that the heavens, sun, and moon move about the earth, which stands still in the center. In England we see similar theological efforts, even after they had become evidently futile. Hutchinson's *Moses's Principia*, Dr. Samuel Pike's *Sacred Philosophy*, the writings of Horne, Bishop Horsley, and President Forbes contain most earnest attacks upon the ideas of Newton, such attacks being based upon Scripture. Dr. John Owen, so famous in the annals of puritanism, declared the Copernican system a "delusive and arbitrary hypothesis, contrary to Scripture"; and even John Wesley declared the new ideas to "tend toward infidelity."

And Protestant peoples were not a whit behind Catholic in following out such teachings. The people of Elbing made themselves merry over a farce in which Copernicus was the main object of ridicule. The people of Nuremberg, a Protestant stronghold, caused a medal to be struck with inscriptions ridiculing the philosopher and his theory.

Why the people at large took this view is easily understandable when we note the attitude of the guardians of learning, both Catholic and Protestant, in that age. It throws great light upon sundry claims by modern theologians to take charge of public instruction and of the evolution of science. So important was it thought to have "sound learning" guarded and "safe science" taught, that in many of the universities, as late as the end of the seventeenth century, professors were forced to take an oath not to hold the "Pythagorean"—that is, the Copernican—idea as to the movement of the heavenly bodies. As the contest went on, professors were forbidden to make known to students the facts revealed by the telescope. Special orders to this effect were issued by the ecclesiastical authorities to the universities and colleges of Pisa, Innsbruck, Louvain, Douay, Salamanca, and others. During generations we find the authorities of these universities boasting that these godless doctrines were kept away from their students. It is touching to hear such boasts made then, just as it is touching now to hear sundry excellent university authorities boast that they discourage the reading of Mill, Spencer, and Darwin. Nor were such attempts to keep the truth from students confined to the Roman Catholic institutions of learning. Strange as it may seem, nowhere were the facts confirming the Copernican theory more carefully kept out of sight than at Wittenberg

—the university of Luther and Melanchthon. About the middle of the sixteenth century there were at that center of Protestant instruction two astronomers of a very high order, Rheticus and Reinhold; both of these, after thorough study, had convinced themselves that the Copernican system was true, but neither of them was allowed to tell this truth to his students. Neither in his lecture announcements nor in his published works did Rheticus venture to make the new system known, and he at last gave up his professorship and left Wittenberg, that he might have freedom to seek and tell the truth. Reinhold was even more wretchedly humiliated. Convinced of the truth of the new theory, he was obliged to advocate the old; if he mentioned the Copernican ideas, he was compelled to overlay them with the Ptolemaic. Even this was not thought safe enough, and in 1571 the subject was intrusted to Peucer. He was eminently "sound," and denounced the Copernican theory in his lectures as "absurd, and unfit to be introduced into the schools."

To clinch anti-scientific ideas more firmly into German Protestant teaching, Rector Hensel wrote a textbook for schools entitled *The Restored Mosaic System of the World,* which showed the Copernican astronomy to be unscriptural.

Doubtless this has a far-off sound; yet its echo comes very near modern Protestantism in the expulsion of Dr. Woodrow by the Presbyterian authorities in South Carolina; the expulsion of Professor Winchell by the Methodist Episcopal authorities in Tennessee; the expulsion of Professor Toy by Baptist authorities in Kentucky; the expulsion of the professors at Beyrout under authority of American Protestant divines—all for holding the doctrines of modern science, and in the last years of the nineteenth century.

But the new truth could not be concealed; it could neither be laughed down nor frowned down. Many minds had received it, but within the hearing of the papacy only one tongue appears to have dared to utter it clearly. This new warrior was that strange mortal, Giordano Bruno. He was hunted from land to land, until at last he turned on his pursuers with fearful invectives. For this he was entrapped at Venice, imprisoned during six years in the dungeons of the Inquisition at Rome, then burned alive, and his ashes scattered to the winds. Still, the new truth lived on. Ten years after the martyrdom of Bruno the truth of Copernicus' doctrine was established by the telescope of Galileo.

Herein was fulfilled one of the most touching of prophecies. Years before, the opponents of Copernicus had said to him, "If your doctrines were true, Venus would show phases like the moon." Copernicus answered: "You are right; I know not what to say; but God is good, and will in time find an answer to this objection." The God-given answer came when, in 1611, the rude telescope of Galileo showed the phases of Venus.

THE WAR UPON GALILEO

On this new champion, Galileo, the whole war was at last concentrated. His discoveries had clearly taken the Copernican theory out of the list of hypotheses, and had placed it before the world as a truth. Against him, then, the war was long and bitter. The supporters of what was called "sound learning" declared his discoveries deceptions and his announcements blasphemy. Semi-scientific professors, endeavoring to curry favor with the Church, attacked him with sham science; earnest preachers attacked him with perverted Scripture; theologians, inquisitors, congregations of cardinals, and at last two popes dealt with him, and, as was supposed, silenced his impious doctrine forever.

I shall present this warfare at some length because, so far as I can find, no careful summary of it has been given in our language, since the whole history was placed in a new light by the revelations of the trial documents in the Vatican Library, honestly published for the first time by L'Epinois in 1867, and since that by Gebler, Berti, Favaro, and others.

The first important attack on Galileo began in 1610, when

he announced that his telescope had revealed the moons of the planet Jupiter. The enemy saw that this took the Copernican theory out of the realm of hypothesis, and they gave battle immediately. They denounced both his method and its results as absurd and impious. As to his method, professors bred in the "safe science" favored by the Church argued that the divinely appointed way of arriving at the truth in astronomy was by theological reasoning on texts of Scripture; and, as to his results, they insisted, first, that Aristotle knew nothing of these new revelations; and, next, that the Bible showed by all applicable types that there could be only seven planets; that this was proved by the seven golden candlesticks of the Apocalypse, by the seven churches of Asia; that from Galileo's doctrine consequences must logically result destructive to Christian truth. Bishops and priests therefore warned their flocks, and multitudes of the faithful besought the Inquisition to deal speedily and sharply with the heretic.

In vain did Galileo try to prove the existence of satellites by showing them to the doubters through his telescope: they either declared it impious to look, or, if they did look, denounced the satellites as illusions from the devil. Good Father Clavius declared that "to see satellites of Jupiter, men had to make an instrument that would create them." In vain did Galileo try to save the great truths he had discovered by his letters to the Benedictine Castelli and the Grand Duchess Christine, in which he argued that literal biblical interpretation should not be applied to science; it was answered that such an argument only made his heresy more detestable; that he was "worse than Luther or Calvin."

The war on the Copernican theory, which up to that time had been carried on quietly, now flamed forth. It was declared that the doctrine was proved false by the standing still of the sun for Joshua, by the declarations that "the foundations of the earth are fixed so firm that they can not be moved," and that the sun "runneth about from one end of the heavens to the other."

But the little telescope of Galileo still swept the heavens, and another revelation was announced—the mountains and valleys in the moon. This brought on another attack. It was declared that this, and the statement that the moon shines by light reflected from the sun, directly contradict the statement in Genesis that the moon is "a great light." To make the matter worse, a painter, placing the moon in a religious picture in its usual position beneath the feet of the Blessed Virgin, outlined on its surface mountains and valleys; this was denounced as a sacrilege logically resulting from the astronomer's heresy.

Still another struggle was aroused when the hated telescope revealed spots upon the sun, and their motion indicating the sun's rotation. Monsignor Elci, head of the University of Pisa, forbade the astronomer Castelli to mention these spots to his students. Father Busaeus, at the University of Innsbruck, forbade the astronomer Scheiner, who had also discovered the spots and proposed a *safe* explanation of them, to allow the new discovery to be known there. At the College of Douay and the University of Louvain this discovery was expressly placed under the ban, and this became the general rule among the Catholic universities and colleges of Europe. The Spanish universities were especially intolerant of this and similar ideas, and up to a recent period their presentation was strictly forbidden in the most important university of all—that of Salamanca.

Such are the consequences of placing the instruction of men's minds in the hands of those mainly absorbed in saving men's souls. Nothing could be more in accordance with the idea recently put forth by sundry ecclesiastics, Catholic and Protestant, that the Church alone is empowered to promulgate scientific truth or direct university instruction. But science gained a victory here also. Observations of the solar spots were reported not only from Galileo in Italy, but from Fabricius in Holland. Father Scheiner then endeavored to make the usual compromise between theology and science.

He promulgated a pseudo-scientific theory, which only provoked derision.

The war became more and more bitter. The Dominican Father Caccini preached a sermon from the text, "Ye men of Galilee, why stand ye gazing up into heaven?" and this wretched pun upon the great astronomer's name ushered in sharper weapons; for, before Caccini ended, he insisted that "geometry is of the devil," and that "mathematicians should be banished as the authors of all heresies." The Church authorities gave Caccini promotion.

Father Lorini proved that Galileo's doctrine was not only heretical but "atheistic," and besought the Inquisition to intervene. The Bishop of Fiesole screamed in rage against the Copernican system, publicly insulted Galileo, and denounced him to the Grand Duke. The Archbishop of Florence solemnly condemned the new doctrines as unscriptural; and Paul V, while petting Galileo and inviting him as the greatest astronomer of the world to visit Rome, was secretly moving the Archbishop of Pisa to pick up evidence against the astronomer.

But by far the most terrible champion who now appeared was Cardinal Bellarmin, one of the greatest theologians the world has known. He was earnest, sincere, and learned, but insisted on making science conform to Scripture. The weapons which men of Bellarmin's stamp used were purely theological. They held up before the world the dreadful consequences which must result to Christian theology were the heavenly bodies proved to revolve about the sun and not about the earth. Their most tremendous dogmatic engine was the statement that "his pretended discovery vitiates the whole Christian plan of salvation." Father Lecazre declared "it casts suspicion on the doctrine of the incarnation." Others declared, "It upsets the whole basis of theology. If the earth is a planet, and only one among several planets, it cannot be that any such great things have been done specially for it as the Christian doctrine teaches. If there are other planets, since God makes nothing in vain, they must be inhabited; but how can their inhabitants be descended from Adam? How can they trace back their origin to Noah's ark? How can they have been redeemed by the Saviour?" Nor was this argument confined to the theologians of the Roman Church; Melanchthon, Protestant as he was, had already used it in his attacks on Copernicus and his school.

In addition to this prodigious theological engine of war there was kept up a fire of smaller artillery in the form of texts and scriptural extracts.

But the war grew still more bitter, and some weapons used in it are worth examining. They are very easily examined, for they are to be found on all the battlefields of science; but on that field they were used with more effect than on almost any other. These weapons are the epithets "infidel" and "atheist." They have been used against almost every man who has ever done anything new for his fellowmen. The list of those who have been denounced as "infidel" and "atheist" includes almost all great men of science, general scholars, inventors, and philanthropists. The purest Christian life, the noblest Christian character, have not availed to shield combatants. Christians like Isaac Newton, Pascal, Locke, Milton, and even Fénelon and Howard, have had this weapon hurled against them. Of all proofs of the existence of a God, those of Descartes have been wrought most thoroughly into the minds of modern men; yet the Protestant theologians of Holland sought to bring him to torture and to death by the charge of atheism, and the Roman Catholic theologians of France thwarted him during his life and prevented any due honors to him after his death.

These epithets can hardly be classed with civilized weapons. They are burning arrows; they set fire to masses of popular prejudice, always obscuring the real question, sometimes destroying the attacking party. They are poisoned weapons. They pierce the hearts of loving women; they alienate dear children; they injure a man after life is ended, for they leave poisoned wounds in the hearts of those who loved him best—fears for his eternal salvation, dread of the Divine wrath upon him. Of course, in these days these weapons, though often effective in vexing good men and in scaring good women, are somewhat blunted; indeed, they not infrequently injure the assailants more than the assailed. It was not so in the days of Galileo; they were then in all their sharpness and venom.

Yet a baser warfare was waged by the Archbishop of Pisa. This man, whose cathedral derives its most enduring fame from Galileo's deduction of a great natural law from the swinging lamp before its altar, was not an archbishop after the noble mould of Borromeo and Fénelon and Cheverus. Sadly enough for the Church and humanity, he was simply a zealot and intriguer: he perfected the plan for entrapping the great astronomer.

Galileo, after his discoveries had been denounced, had written to his friend Castelli and to the Grand Duchess Christine two letters to show that his discoveries might be reconciled with Scripture. On a hint from the Inquisition at Rome, the archbishop sought to get hold of these letters and exhibit them as proofs that Galileo had uttered heretical views of theology and of Scripture, and thus to bring him into the clutches of the Inquisition. The archbishop begs Castelli, therefore, to let him see the original letter in the handwriting of Galileo. Castelli declines. The archbishop then, while, as is now revealed, writing constantly and bitterly to the Inquisition against Galileo, professes to Castelli the greatest admiration of Galileo's genius and a sincere desire to know more of his discoveries. This not succeeding, the archbishop at last throws off the mask and resorts to open attack.

The whole struggle to crush Galileo and to save him would be amusing were it not so fraught with evil. There were intrigues and counter-intrigues, plots and counter-plots, lying and spying; and in the thickest of this seething, squabbling, screaming mass of priests, bishops, archbishops, and cardinals, appear two popes, Paul V and Urban VIII. It is most suggestive to see in this crisis of the Church, at the tomb of the prince of the apostles, on the eve of the greatest errors in Church policy the world has known, in all the intrigues and deliberations of these consecrated leaders of the Church, no more evidence of the guidance or presence of the Holy Spirit than in a caucus of New York politicians at Tammany Hall.

But the opposing powers were too strong. In 1615 Galileo was summoned before the Inquisition at Rome, and the mine which had been so long preparing was sprung. Sundry theologians of the Inquisition having been ordered to examine two propositions which had been extracted from Galileo's letters on the solar spots, solemnly considered these points during about a month and rendered their unanimous decision as follows: *"The first proposition, that the sun is the center and does not revolve about the earth, is foolish, absurd, false in theology, and heretical, because expressly contrary to Holy Scripture"*; and *"the second proposition, that the earth is not the center but revolves about the sun, is absurd, false in philosophy, and, from a theological point of view at least, opposed to the true faith."*

The Pope himself, Paul V, now intervened again: he ordered that Galileo be brought before the Inquisition. Then the greatest man of science in that age was brought face to face with the greatest theologian—Galileo was confronted by Bellarmin. Bellarmin shows Galileo the error of his opinion and orders him to renounce it. De Lauda, fortified by a letter from the Pope, gives orders that the astronomer be placed in the dungeons of the Inquisition should he refuse to yield. Bellarmin now commands Galileo, "in the name of His Holiness the Pope and the whole Congregation of the Holy Office, to relinquish altogether the opinion that the sun is the center of the world and immovable, and that the earth moves, nor henceforth to hold, teach, or defend it in any way whatsoever, verbally or in writing." This injunction Galileo acquiesces in and promises to obey.

This was on the 26th of February, 1616. About a fortnight later the Congregation of the Index, moved thereto, as the letters and documents now brought to light show, by Pope Paul V, solemnly rendered a decree that *"the doctrine of the double motion of the earth about its axis and about the sun is false, and entirely contrary to Holy Scripture";* and that this opinion must neither be taught nor advocated. The same decree condemned all writings of Copernicus and *"all writings which affirm the motion of the earth."* The great work of Copernicus was interdicted until corrected in accordance with the views of the Inquisition; and the works of Galileo and Kepler, though not mentioned by name at that time, were included among those implicitly condemned as "affirming the motion of the earth."

The condemnations were inscribed upon the *Index;* and, finally, the papacy committed itself as an infallible judge and teacher to the world by prefixing to the *Index* the usual papal bull giving its monitions the most solemn papal sanction. To teach or even read the works denounced or passages condemned was to risk persecution in this world and damnation in the next. Science had apparently lost the decisive battle.

For a time after this judgment Galileo remained in Rome, apparently hoping to find some way out of this difficulty; but he soon discovered the hollowness of the protestations made to him by ecclesiastics, and, being recalled to Florence, remained in his hermitage near the city in silence, working steadily, indeed, but not publishing anything save by private letters to friends in various parts of Europe.

But at last a better vista seemed to open for him. Cardinal Barberini, who had seemed liberal and friendly, became pope under the name of Urban VIII. Galileo at this conceived new hopes, and allowed his continued allegiance to the Copernican system to be known. New troubles ensued.

Galileo was induced to visit Rome again, and Pope Urban tried to cajole him into silence, personally taking the trouble to show him his errors by argument. Other opponents were less considerate, for works appeared attacking his ideas—works all the more unmanly, since their authors knew that Galileo was restrained by force from defending himself. Then, too, as if to accumulate proofs of the unfitness of the Church to take charge of advanced instruction, his salary as a professor at the University of Pisa was taken from him, and sapping and mining began. Just as the Archbishop of Pisa some years before had tried to betray him with honeyed words to the Inquisition, so now Father Grassi tried it, and, after various attempts to draw him out by flattery, suddenly denounced his scientific ideas as "leading to a denial of the Real Presence in the Eucharist."

For the final assault upon him a park of heavy artillery was at last wheeled into place. It may be seen on all the scientific battlefields. It consists of general denunciation; and in 1631 Father Melchior Inchofer, of the Jesuits, brought his artillery to bear upon Galileo with this declaration: "The opinion of the earth's motion is of all heresies the most abominable, the most pernicious, the most scandalous; the immovability of the earth is thrice sacred; argument against the immortality of the soul, the existence of God, and the incarnation, should be tolerated sooner than an argument to prove that the earth moves."

From the other end of Europe came a powerful echo. From the shadow of the Cathedral of Antwerp, the noted theologian Fromundus gave forth his famous treatise, the *Ant-Aristarchus*. Its very title page was a contemptuous insult to the memory of Copernicus since it paraded the assumption that the new truth was only an exploded theory of a pagan astronomer. Fromundus declares that "sacred Scripture fights against the Copernicans." To prove that the sun revolves about the earth, he cites the passage in the Psalms which speaks of the sun "which cometh forth as a bridegroom out of his chamber." To prove that the earth stands still, he quotes a passage from Ecclesiastes, "The earth standeth fast forever." To show the utter futility of the Copernican theory, he declares that, if it were true, "the wind would constantly blow from the east"; and that "buildings and the earth itself would fly off with such a rapid motion that men would have to be provided with claws like cats to enable them to hold fast to the earth's surface." Greatest weapon of them all, he works up, by the use of Aristotle and St. Thomas Aquinas, a demonstration from theology and science combined, that the earth *must* stand in the center, and that the sun *must* revolve about it. Nor was it merely fanatics who opposed the truth revealed by Copernicus; such strong men as Jean Bodin, in France, and Sir Thomas Browne, in England, declared against it as evidently contrary to Holy Scripture.

VICTORY OF THE CHURCH OVER GALILEO

While news of triumphant attacks upon him and upon the truth he had established were coming in from all parts of Europe, Galileo prepared a careful treatise in the form of a dialogue, exhibiting the argument for and against the Copernican and Ptolemaic systems, and offered to submit to any conditions that the Church tribunals might impose, if they would allow it to be printed. At last, after discussions which extended through eight years, they consented, imposing a humiliating condition—a preface written in accordance with the ideas of Father Ricciardi, Master of the Sacred Palace, and signed by Galileo, in which the Copernican theory was virtually exhibited as a play of the imagination, and not at all as opposed to the Ptolemaic doctrine reasserted in 1616 by the Inquisition under the direction of Pope Paul V.

This new work of Galileo—the *Dialogo*—appeared in 1632 and met with prodigious success. It put new weapons into the hands of the supporters of the Copernican theory. The pious preface was laughed at from one end of Europe to the other. This roused the enemy; the Jesuits, Dominicans, and the great majority of the clergy returned to the attack more violently than ever, and in the midst of them stood Pope Urban VIII, most bitter of all. His whole power was now thrown against Galileo. He was touched in two points: first, in his personal vanity, for Galileo had put the Pope's arguments into the mouth of one of the persons in the dialogue and their refutation into the mouth of another; but, above all, he was touched in his religious feelings. Again and again His Holiness insisted to all comers on the absolute and specific declarations of Holy Scripture, which prove that the sun and heavenly bodies revolve about the earth, and declared that to gainsay them is simply to dispute revelation. Certainly, if one ecclesiastic more than another ever seemed *not* under the care of the Spirit of Truth in all this matter, it was Urban VIII.

Herein was one of the greatest pieces of ill fortune that has ever befallen the older Church. Had Pope Urban been broad-minded and tolerant like Benedict XIV, or had he been taught moderation by adversity like Pius VII, or had he possessed the large scholarly qualities of Leo XIII, now reigning, the vast scandal of the Galileo case would never have burdened the Church: instead of devising endless quibbles and special pleadings to escape responsibility for this colossal blunder, its defenders could have claimed forever for the Church the glory of fearlessly initiating a great epoch in human thought.

But it was not so to be. Urban was not merely Pope; he was also a prince of the house of Barbarini, and therefore doubly angry that his arguments had been publicly controverted.

The opening strategy of Galileo's enemies was to forbid the sale of his work; but this was soon seen to be unavailing, for the first edition had already been spread throughout

Europe. Urban now became more angry than ever, and both Galileo and his works were placed in the hands of the Inquisition. In vain did the good Benedictine Castelli urge that Galileo was entirely respectful to the Church; in vain did he insist, "nothing that can be done now will hinder the earth from revolving." He was dismissed in disgrace, and Galileo was forced to appear in the presence of the dread tribunal without defender or adviser. There, as was so long concealed, but as is now fully revealed, he was menaced with torture again and again by express order of Pope Urban, and, as is also thoroughly established from the trial documents themselves, forced to abjure under threats, and subjected to imprisonment by command of the Pope; the Inquisition deferring in this whole matter to the papal authority. All the long series of attempts made in the supposed interest of the Church to mystify these transactions have at last failed. The world knows now that Galileo was subjected certainly to indignity, to imprisonment, and to threats equivalent to torture, and was at last forced to pronounce publicly and on his knees his recantation, as follows:

"I, Galileo, being in my seventieth year, being a prisoner and on my knees, and before your Eminences, having before my eyes the Holy Gospel, which I touch with my hands, abjure, curse, and detest the error and the heresy of the movement of the earth."

He was vanquished indeed, for he had been forced, in the face of all coming ages, to perjure himself. To complete his dishonor, he was obliged to swear that he would denounce to the Inquisition any other man of science whom he should discover to be supporting the "heresy of the motion of the earth."

Many have wondered at this abjuration, and on account of it have denied to Galileo the title of martyr. But let such gainsayers consider the circumstances. Here was an old man —one who had reached the allotted threescore years and ten —broken with disappointments, worn out with labors and cares, dragged from Florence to Rome, with the threat from the Pope himself that if he delayed he should be "brought in chains"; sick in body and mind, given over to his oppressors by the Grand Duke who ought to have protected him, and on his arrival in Rome threatened with torture. What the Inquisition was he knew well. He could remember as but of yesterday the burning of Giordano Bruno in that same city for scientific and philosophic heresy; he could remember, too, that only eight years before this very time De Dominis, Archbishop of Spalatro, having been seized by the Inquisition for scientific and other heresies, had died in a dungeon, and that his body and his writings had been publicly burned.

To the end of his life—nay, after his life was ended—the persecution of Galileo was continued. He was kept in exile from his family, from his friends, from his noble employments, and was held rigidly to his promise not to speak of his theory. When, in the midst of intense bodily sufferings from disease, and mental sufferings from calamities in his family, he besought some little liberty, he was met with threats of committal to a dungeon. When, at last, a special commission had reported to the ecclesiastical authorities that he had become blind and wasted with disease and sorrow, he was allowed a little more liberty, but that little was hampered by close surveillance. He was forced to bear contemptible attacks on himself and on his works in silence; to see the men who had befriended him severely punished; Father Castelli banished; Ricciardi, the Master of the Sacred Palace, and Ciampoli, the papal secretary, thrown out of their positions by Pope Urban, and the Inquisitor at Florence reprimanded for having given permission to print Galileo's work. He lived to see the truths he had established carefully weeded out from all the Church colleges and universities in Europe; and, when in a scientific work he happened to be spoken of as "renowned," the Inquisition ordered the substitution of the word "notorious."

The Judgment of Posterity

[When White published his *History of the Warfare of Science with Theology in Christendom* in 1896, the Galileo "case" seemed to have been thoroughly closed; the scientific world seemed to have rendered a final judgment settling the matter: however mistaken or tactless he may have been religiously, morally, sociologically or historically, Galileo was certainly right scientifically when he asserted that the earth turns on its axis and circles around the sun, and his opponents were as certainly wrong.

And then suddenly a new current of scientific thought made its way into prominence. Great scientists, masters of the scientific method developed by Galileo and Newton, by using that very method, worked a total revolution in mathematical-empirical science upsetting the very foundations of the traditional Galilean-Newtonian physics. The central figure in this revolution was, of course, Albert Einstein, to whom the readers of this book need no introduction.

Einstein was tireless in his praise of Galileo and Newton as pioneers in the development of science. Most emphatically he upheld the right of Galileo to be honored as the "father of modern physics—indeed, of modern science altogether" [see Introduction, p. 10]. Nevertheless, it was soon recognized by serious students that Einstein's own scientific achievement, especially his later theory of relativity, could not but lead, as Professor Morris R. Cohen has expressed it, to a reopening of "the issue between Galileo and those who condemned him for saying that the earth *is* in motion."

In the following selection Einstein very soberly assesses the achievement of Galileo, remarking only in passing on the "relativity" of knowledge acquired by use of the method perfected by Galileo, without pressing the point. The point is pressed, however, in the two selections by Morris Cohen and Philipp Frank that follow.]

Foreword*

by

ALBERT EINSTEIN

Galileo's *Dialogue Concerning the Two Chief World Systems* [translated elsewhere in our text as the *Dialogue on the Great World Systems*] is a mine of information for anyone interested in the cultural history of the Western world and its influence upon economic and political development.
A man is here revealed who possesses the passionate will, the intelligence, and the courage to stand up as the representative of rational thinking against the host of those who, relying on the ignorance of the people and the indolence of teachers in priest's and scholar's garb, maintain and defend their positions of authority. His unusual literary gift enables him to address the educated men of his age in such clear and impressive language as to overcome the anthropocentric and mythical thinking of his contemporaries and to lead them back to an objective and causal attitude toward the cosmos, an attitude which had become lost to humanity with the decline of Greek culture.
In speaking this way I notice that I, too, am falling in with the general weakness of those who, intoxicated with devotion, exaggerate the stature of their heroes. It may well be that during the seventeenth century the paralysis of mind brought about by the rigid authoritarian tradition of the

* Galileo, *Dialogue Concerning the Two Chief World Systems— Ptolemaic & Copernican.* Foreword by Albert Einstein, translated by Sonja Bargmann. Reprinted by permission of the University of California Press.

Dark Ages had already so far abated that the fetters of an obsolete intellectual tradition could not have held much longer—with or without Galileo.

Yet these doubts concern only a particular case of the general problem concerning the extent to which the course of history can be decisively influenced by single individuals whose qualities impress us as accidental and unique. As is understandable, our age takes a more sceptical view of the role of the individual than did the eighteenth and the first half of the nineteenth century. For the extensive specialization of the professions and of knowledge lets the individual appear "replaceable," as it were, like a part of a mass-produced machine.

Fortunately, our appreciation of the *Dialogue* as a historical document does not depend upon our attitude toward such precarious questions. To begin with, the *Dialogue* gives an extremely lively and persuasive exposition of the then prevailing views on the structure of the cosmos in the large. The naïve picture of the earth as a flat disc, combined with obscure ideas about star-filled space and the motions of the celestial bodies, prevalent in the early Middle Ages, represented a deterioration of the much earlier conceptions of the Greeks, and in particular of Aristotle's ideas and of Ptolemy's consistent spatial concept of the celestial bodies and their motions. The conception of the world still prevailing at Galileo's time may be described as follows:

There is space, and within it there is a preferred point, the center of the universe. Matter—at least its denser portion —tends to approach this point as closely as possible. Consequently, matter has assumed approximately spherical shape (earth). Owing to this formation of the earth the center of the terrestrial sphere practically coincides with that of the universe. Sun, moon, and stars are prevented from falling toward the center of the universe by being fastened onto rigid (transparent) spherical shells whose centers are identical with that of the universe (or space). These spherical shells revolve around the immovable globe (or center of the universe) with slightly differing angular velocities. The lunar shell has the smallest radius; it encloses everything "terrestrial." The outer shells with their heavenly bodies represent the "celestial sphere" whose objects are envisaged as eternal, indestructible, and inalterable, in contrast to the "lower, terrestrial sphere" which is enclosed by the lunar shell and contains everything that is transitory, perishable, and "corruptible."

Naturally, this naïve picture cannot be blamed on the Greek astronomers who, in representing the motions of the celestial bodies, used abstract geometrical constructions which grew more and more complicated with the increasing precision of astronomical observations. Lacking a theory of mechanics they tried to reduce all complicated (apparent) motions to the simplest motions they could conceive, namely, uniform circular motions and superpositions thereof. Attachment to the idea of circular motion as the truly natural one is still clearly discernible in Galileo; probably it is responsible for the fact that he did not *fully* recognize the law of inertia and its fundamental significance.

Thus, briefly, had the ideas of later Greece been crudely adapted to the barbarian, primitive mentality of the Europeans of that time. Though not causal, those Hellenistic ideas had nevertheless been objective and free from animistic views—a merit which, however, can be only conditionally conceded to Aristotelian cosmology.

In advocating and fighting for the Copernican theory Galileo was not only motivated by a striving to simplify the representation of the celestial motions. His aim was to substitute for a petrified and barren system of ideas the unbiased and strenuous quest for a deeper and more consistent comprehension of the physical and astronomical facts.

The form of the dialogue used in his work may be partly due to Plato's shining example; it enabled Galileo to apply his extraordinary literary talent to the sharp and vivid confrontation of opinions. To be sure, he wanted to avoid an open commitment of these controversial questions that would have delivered him to destruction by the Inquisition. Galileo had, in fact, been expressly forbidden to advocate the Copernican theory. Apart from its revolutionary factual content the *Dialogue* represents a downright roguish attempt to comply with this order in appearance and yet in fact to disregard it. Unfortunately, it turned out that the Holy Inquisition was unable to appreciate adequately such subtle humor.

The theory of the immovable earth was based on the hypothesis that an abstract center of the universe exists. Supposedly, this center causes the fall of heavy bodies at the earth's surface, since material bodies have the tendency to approach the center of the universe as far as the earth's impenetrability permits. This leads to the approximately spherical shape of the earth.

Galileo opposes the introduction of this "nothing" (center of the universe) that is yet supposed to act on material bodies; he considers this quite unsatisfactory.

But he also draws attention to the fact that this unsatisfactory hypothesis accomplishes too little. Although it accounts for the spherical shape of the earth it does not explain the spherical shape of the other heavenly bodies. However, the lunar phases and the phases of Venus, which latter he had discovered with the newly invented telescope, proved the spherical shape of these two celestial bodies; and the detailed observations of the sun-spots proved the same for the sun. Actually, at Galileo's time there was hardly any doubt left as to the spherical shape of the planets and stars.

Therefore, the hypothesis of the "center of the universe" had to be replaced by one which would explain the spherical shape of the stars, and not only that of the earth. Galileo says quite clearly that there must exist some kind of interaction (tendency to mutual approach) of the matter constituting a star. The same cause has to be responsible (after relinquishing the "center of the universe") for the free fall of heavy bodies at the earth's surface.

Let me interpolate here that a close analogy exists between Galileo's rejection of the hypothesis of a center of the universe for the explanation of the fall of heavy bodies, and the rejection of the hypothesis of an inertial system for the explanation of the inertial behavior of matter. (The latter is the basis of the theory of general relativity.) Common to both hypotheses is the introduction of a conceptual object with the following properties:

(1) It is not assumed to be real, like ponderable matter (or a "field").

(2) It determines the behavior of real objects, but it is in no way affected by them.

The introduction of such conceptual elements, though not exactly inadmissible from a purely logical point of view, is repugnant to the scientific instinct.

Galileo also recognized that the effect of gravity on freely falling bodies manifests itself in a vertical acceleration of constant value; likewise that an unaccelerated horizontal motion can be superposed on this vertical accelerated motion.

These discoveries contain essentially—at least qualitatively—the basis of the theory later formulated by Newton. But first of all the general formulation of the principle of inertia is lacking, although this would have been easy to obtain from Galileo's law of falling bodies by a limiting process. (Transition to vanishing vertical acceleration.) Lacking also is the idea that the same matter which causes a vertical acceleration at the surface of a heavenly body can also accelerate another heavenly body; and that such acceleration together with inertia can produce revolving motions. There was achieved, however, the knowledge that the presence of matter (earth) causes an acceleration of free bodies (at the surface of the earth).

It is difficult for us today to appreciate the imaginative power made manifest in the precise formulation of the con-

cept of acceleration and in the recognition of its physical
significance.

Once the conception of the center of the universe had,
with good reason, been rejected, the idea of the immovable
earth, and, generally, of an exceptional role of the earth,
was deprived of its justification. The question of what, in
describing the motion of heavenly bodies, should be considered "at rest" became thus a question of convenience.
Following Aristarchus and Copernicus, the advantages of
assuming the sun to be at rest are set forth (according to
Galileo not a pure convention but a hypothesis which is
either "true" or "false"). Naturally, it is argued that it is
simpler to assume a rotation of the earth around its axis
than a common revolution of all fixed stars around the earth.
Furthermore, the assumption of a revolution of the earth
around the sun makes the motions of the inner and outer
planets appear similar and does away with the troublesome
retrograde motions of the outer planets, or rather explains
them by the motion of the earth around the sun.

Convincing as these arguments may be—in particular
coupled with the circumstance, detected by Galileo, that
Jupiter with its moons represents so to speak a Copernican
system in miniature—they still are only of a qualitative nature.
For since we human beings are tied to the earth, our observations will never directly reveal to us the "true" planetary motions, but only the intersections of the lines of sight
(earth-planet) with the "fixed-star sphere." A support of the
Copernican system over and above qualitative arguments
was possible only by determining the "true orbits" of the
planets—a problem of almost insurmountable difficulty, which,
however, was solved by Kepler (during Galileo's lifetime) in a
truly ingenious fashion. But this decisive progress did not leave
any traces in Galileo's life work—a grotesque illustration of the
fact that creative individuals are often not receptive.

Galileo takes great pains to demonstrate that the hypothesis of the rotation and revolution of the earth is not refuted

by the fact that we do not observe any mechanical effects of these motions. Strictly speaking, such a demonstration was impossible because a complete theory of mechanics was lacking. I think it is just in the struggle with this problem that Galileo's originality is demonstrated with particular force. Galileo is, of course, also concerned to show that the fixed stars are too remote for parallaxes produced by the yearly motion of the earth to be detectable with the measuring instruments of his time. This investigation also is ingenious, notwithstanding its primitiveness.

It was Galileo's longing for a mechanical proof of the motion of the earth which misled him into formulating a wrong theory of the tides. The fascinating arguments in the last conversation would hardly have been accepted as proofs by Galileo, had his temperament not got the better of him. It is hard for me to resist the temptation to deal with this subject more fully.

The *leitmotif* which I recognize in Galileo's work is the passionate fight against any kind of dogma based on authority. Only experience and careful reflection are accepted by him as criteria of truth. Nowadays it is hard for us to grasp how sinister and revolutionary such an attitude appeared at Galileo's time, when merely to doubt the truth of opinions which had no basis but authority was considered a capital crime and punished accordingly. Actually we are by no means so far removed from such a situation even today as many of us would like to flatter ourselves; but in theory, at least, the principle of unbiased thought has won out, and most people are willing to pay lip service to this principle.

It has often been maintained that Galileo became the father of modern science by replacing the speculative, deductive method with the empirical, experimental method. I believe, however, that this interpretation would not stand close scrutiny. There is no empirical method without speculative concepts and systems; and there is no speculative thinking whose concepts do not reveal, on closer investigation, the empirical material from which they stem. To put into sharp contrast the empirical and the deductive attitude is misleading, and was entirely foreign to Galileo. Actually it was not until the nineteenth century that logical (mathematical) systems whose structures were completely independent of any empirical content had been cleanly extracted. Moreover, the experimental methods at Galileo's disposal were so imperfect that only the boldest speculation could possibly bridge the gaps between the empirical data. (For example, there existed no means to measure times shorter than a second.) The antithesis Empiricism *vs.* Rationalism does not appear as a controversial point in Galileo's work. Galileo opposes the deductive methods of Aristotle and his adherents only when he considers their premises arbitrary or untenable, and he does not rebuke his opponents for the mere fact of using deductive methods. In the first dialogue, he emphasizes in several passages that according to Aristotle, too, even the most plausible deduction must be put aside if it is incompatible with empirical findings. And on the other hand, Galileo himself makes considerable use of logical deduction. His endeavors are not so much directed at "factual knowledge" as at "comprehension." But to comprehend is essentially to draw conclusions from an already accepted logical system.

[EDITOR'S NOTE]

This English translation originally appeared with the German text facing it, thus enabling careful readers to distinguish, if they would, between the literal and the translated sense of Einstein's words. One conspicuous liberty which the translator took deserves to be noted as a reminder that preconceived ideas, long accepted on authority, are as apt now as they were in Galileo's time to weigh heavily upon human judgment. Making a modest concession to relatively recent scholarship, Einstein, in a brief paragraph of this Foreword, lends some of the weight of his authority to discredit, in part, one of the old fallacies that originated in the historical ignorance of seventeenth- and eighteenth-century rationalists regarding the state of cosmological knowledge in the Middle Ages. "Die im früheren Mittelalter herrschende kindliche Auffassung der Erde als eine flachen Scheibe, verknüpft mit ganz unklaren Ideen über den von den Sternen erfüllten Raum und die Bewegung der Gestirne, waren längst durch das Weltbild der Griechen, speziell durch Ideen des Aristoteles und durch die ptolemäische konsequente räumliche Auffassung der Gestirne und deren Bewegung verbessert." Einstein here reminds his readers that the childish notions allegedly entertained in the early Middle Ages about the stars and their motions, and about the earth's being a flat disc, *were finally corrected* ("waren längst . . . verbessert") by the medieval study of Greek cosmology, and especially by the study of Aristotelian and Ptolomaic thought. The English version, refusing to allow Einstein to say "waren längst . . . verbessert," reads: "The naïve picture of the earth as a flat disc, combined with obscure ideas about star-filled space and the motions of the celestial bodies, prevalent in the early Middle Ages, represented a deterioration of the much earlier conceptions of the Greeks, and in particular of Aristotle's ideas and of Ptolemy's consistent spatial concept of the celestial bodies and their motions." Like those contemporaries of Galileo who reportedly refused to believe that what they could have seen with their own eyes looking through the telescope, the translator here refuses to believe what is readily visible in Einstein's text, insisting that Einstein himself must be made to conform with the sacred doctrines of the prevailing popular "authorities" who have long been teaching, out of the bounty of their historical ignorance, that, before the time of Columbus, if not before that of Galileo, all Christendom believed that the earth was flat. The best modern scholarship, indeed, gives no warrant to say that it was ever acceptable in the Middle Ages, late or early, to pretend that the world was a flat disc.

from

Studies in Philosophy and Science*

by

MORRIS R. COHEN

[Morris R. Cohen, who taught for many years at the City College of New York, and at the University of Chicago, brought to his study of the implications of modern science a vast culture embracing the fields of law and philosophy as well as the history of science. In the reading presented here he strives to make explicit what was implicit in the Einsteinian revolution which upset the traditional confidence in the foundations of the Galilean-Newtonian mechanics, and he is eminently successful in his endeavor.]

* Morris R. Cohen, *Studies in Philosophy and Science*. Reprinted by permission of Henry Holt and Company, Inc.

The Law of Gravitation and the More General Theory of Relativity

(Taken from the Chapter "EINSTEIN'S THEORY OF RELATIVITY")

For over two centuries Newton's law of gravitation has served as the model or stock example of a law of nature. All efforts at scientific truth, even in the undeveloped social sciences, have regarded the discovery of similar laws as the ideal of scientific attainment. Any attempt, therefore, such as Einstein's, to modify and improve upon Newton's law must be viewed as having more than a merely technical interest.

The belief in simple and eternal laws of nature back of the persistent irregularity and instability of sensible phenomena, grew out of the ancient Neo-Platonic tradition that to the mind that approaches divine insight the book of nature is written in simple geometric lines. All the great founders of modern science, Copernicus, Kepler, Galileo, Descartes, and Newton shared this faith. The splendid results which followed their search for simple laws gave their faith the unique position of being the only one to have almost completely escaped serious assaults from the modern critical spirit. For despite their professions of welcome to anyone who can challenge their first principles, philosophers and scientists are made of the same human clay as theologians and lawyers or men of affairs, and have the same organic aversion for the thought which disturbs established and comfortable certainties. But many a faith that has been unassailable by direct frontal attack has been forced to yield or to reorganize by pressure from other quarters; and the faith in simple eternal laws of nature has in fact been undermined on the experimental side by the progressive improvement of our instruments of measurement, and on the mathematical side by the discovery of non-Euclidean geometry. The former has led to the view that our seemingly absolute laws of nature are but the statistical averages of the behavior of large numbers of inherently variable elements, and reflection on non-Euclidean geometry has pressed forward the thought that many diverse accounts of our fragmentary experience of the physical world can all claim to be equally true.

Everyone who has ever worked in a laboratory or with instruments of precision knows that the simple laws of nature, so clearly formulated in elementary and popular treatises, are never verified with absolute accuracy. The results of actual measurements always differ. We attribute this universal discrepancy between our theoretic formulae and our actual measurements not to our theory but to the "error" of our instruments. But the fact is that the refinement or improvement of our instruments never eliminates this discrepancy. On the contrary it often compels us to abandon the simple law in favor of a more complicated one. Boyle's law of the simple inverse proportionality between the volume and the pressure of gases has now yielded to the more complicated equation of Van de Waals; and the fate of Coulomb's law in electricity has indicated that the similarly formulated law of gravitation might also show itself to be but a first approximation in need of correction as our knowledge becomes more accurate.

That the acceptance of the theory of relativity involves some modification of the Newtonian theory of gravitation, and indeed of the whole Newtonian mechanics, is obvious from at least two considerations. The Newtonian mechanics is based on the assumption of the constancy of mass (popularly known as "the indestructibility of matter"), but from

THE ACHIEVEMENT OF GALILEO

the theory of relativity it necessarily follows that the mass of a body varies with its velocity, and is different in the direction of its motion than in any direction perpendicular to it. Again, according to Newton the force of gravity is transmitted instantaneously or practically so, whereas according to the relativity theory there can be no greater velocity than that of light. According to the Newtonian mechanics the gravity or weight of a body is proportional to its mass or inertia, and the latter is a constant and independent constituent of energy. Modern experiments have suggested that possibly what we call mass is itself of electromagnetic origin; or, at any rate, that radiant energy like light and cathode rays offer inertia or resistance to change which may well be called mass, and that such mass varies with its velocity. Thus considerations of experimental physics as well as deductions from the theory of relativity led Einstein, soon after publishing his paper of 1905, to the belief that energy itself has inertia or mass, and, therefore, gravity, and later to the hypothesis that gravitation depends not only on mass and distance but on other factors as well. Maxwell had already shown that light must exert pressure. It was natural for Einstein to take the next step and show that light must also have gravity. But the universal assumption that light travels in straight lines, and the difficulty of finding experimental tests for his theory, offered seemingly insuperable difficulties. To overcome these difficulties Einstein resorted to non-Euclidean geometry, made use of new mathematical methods, and widened or generalized his original theory of relativity.

In the halls of fame the names of Lobachevski and Riemann, the discoverers of non-Euclidean geometry, may seldom be heard. Riemann died in the prime of youth, and the imaginative genius of Lobachevski was smothered by the bleak prison doors of the remote and unenlightened University of Kazan. Yet these two men initiated one of the greatest revolutions in the history of human thought—

The Judgment of Posterity

they undermined for all time the unquestioned sanctity of axioms or first principles. For over two thousand years Euclid's geometry had served as the model for all science, philosophy, and theology. It was universally taken for granted—and most people still assume—that in every field there are axioms or first principles that cannot be doubted because they are self-evident, i.e., simple, clear and conclusive on simple inspection.

Lobachevski showed that one of Euclid's axioms, that relating to parallel lines, could well be questioned: and Riemann went further in questioning the assumption (for that is what every axiom really is), that through any two points only one straight line can be drawn.

Now though Euclidean geometry is still, because of its relative simplicity, the most convenient for ordinary lengths and areas, there is no mathematical or physical reason against the attempt to describe the astronomic universe in terms of one of these other geometries. It depends upon our choice as to what shall be the physical test of a straight line. If the captain of a ship defines a straight line on the surface of the earth or sea as the shortest distance between two points, he has in fact chosen the Riemannian geometry, since between two poles of the terrestrial sphere any number of such straight lines can be drawn. Similarly Einstein may with very good reason take the path of a light ray as the test of straightness. If in addition to this he also holds that light, like a projectile, proceeds not only under its own energy, but is deflected by a gravitational field, such deflection does not contradict the original definition of straightness, but only compels the use of non-Euclidean geometry.

Such a procedure may appear arbitrary to those who dislike all departure from the usual ways of doing things; and doubtless it is so. Only it must not be forgotten that the usual procedure is also arbitrary. Indeed, so long as man's knowledge of the universe is fragmentary, every attempt to formulate its nature must contain arbitrary elements. Arbitrary procedures, however, are justified if they lead to significant discoveries and in this respect Einstein's method is certainly justified.

In developing his theory of gravitation Einstein came into conflict with his own original theory of relativity, which he might also have called the theory of the absolute constancy of the velocity of light. Now if a gravitational field affects the path of light it cannot, according to Einstein's mathematics, leave its velocity unaffected. In his original theory he had shown that our ordinary units of time and distance were variable in relation to the constant velocity of light. What now is the constant with reference to which the velocity of light varies? The answer to this is that while theoretically the velocity of light can be constant only in the absence of marked gravitational influence, the mass of the earth is practically negligible in cosmic relations in which the sun and the "fixed" stars enter. Hence the original theory of relativity may be regarded as approximately true on the surface of the earth or wherever gravity may be viewed as a non-disturbing factor with reference to light.

To speak of the deflection of light by a gravitational field may seem to involve the old view of gravity as a force which pulls things together. This is not Einstein's view. His theory aims to be purely descriptive, and gravity appears in it not as a force, but rather as a property of a space or field. All we know of gravitation is that in certain portions of space all bodies, no matter what their constitution, are uniformly accelerated. Indeed, the phenomenon of deflected light would result in precisely the same way if there were no such thing as gravitation, but if the observer were moving with accelerated velocity in a direction perpendicular to the path of a light ray. The reader can make this clear to himself by imagining himself in an elevator going down with a uniform acceleration equal to that of a freely falling body. In such an elevator no free object can fall to the floor and a horizontally shot projectile, which, as seen from the earth, falls in a curved line, would here describe a perfectly horizontal line. Conversely if the path of the projectile or a light ray be perfectly horizontal to an observer on the earth, it will be curved to an observer in the elevator.

Considerations such as these have led Einstein to generalize his original theory of relativity. Instead of saying that the laws of nature are the same whether we suppose the observer to be at rest or in uniform motion, he now says the laws of nature are the same whether we suppose the observer to be at rest or in any kind of motion, accelerated or rotatory.

To realize something of the meaning of this statement, let the reader imagine a group of unusually gifted scientific observers confined since birth by some mysterious fate in a well-supplied Pullman car, and unable to learn anything of the outside world except by means of the light rays which stream in through their windows. (If this sounds too fanciful, let the reader remember that our earth is just such a car.) If such scientific observers begin to formulate the laws of nature they will naturally suppose their car to be at rest and all other things in motion in diverse ways. Their laws or equations of motions would be inordinately more complex than those familiar to us, if the car did not always move with uniform velocity. Imagine one of our scientists, as gifted as Copernicus or the early Pythagoreans, saying to himself, "Why not suppose that the earth outside of my windows is at rest and that my car is in motion?" If he did, he would be able to simplify his account of nature enormously. The sudden lurch forward or backward of loose objects, for instance, would be explained not in some, to us, mysterious and complicated way, but by the inertia of things in motion. If now our scientist exultingly claimed before his fellow passengers that this proved that their car was really in motion, he might be stopped by one of them having the genius of Einstein and admonished as follows: "Hold on! You have undoubtedly discovered a new and simpler system of laws or equations to describe the course of nature. But what right have you to claim that your account is truer than the one which we have always hitherto used? Do you suppose that nature has no other care but to conduct herself in such a way as to make it possible for us to describe her conduct in simple laws? Besides I can show you a system of equations by which you can pass from every proposition in your old account of nature to a corresponding proposition in your new account."

The reader who knows something of the history of science will recognize that our example shows Einstein's later theory of relativity as reopening the issue between Galileo and those who condemned him for saying that the earth *is* in motion. If there is no unique absolute space and all motion is relative, it is just as true to say that the earth moves with reference to the car as to say that the car moves with reference to the earth. With our fixed limits of conception and expression, it may be extremely inconvenient or ridiculous to say that every time we drop an object the earth moves up to it; but it would be difficult to prove the falsity of this way of putting it. Similarly with regard to the revolution of the earth around its axis, which Einstein after the example of Mach, calls a revolution with reference to the "fixed" stars. It would be vain to repeat against Einstein the old arguments for the absolute rotation of the earth, based on Foucault's pendulum or the bulging of the earth at the equator. He shows that it is possible to define a space with regard to which the fixed stars are rotating. In such a space the earth may be considered at rest, and the phenomena which in Newtonian mechanics are called gravitational and centrifugal would change places. Since both are proportional to the mass of the earth there would be no experimental difference. Notice that Einstein does not justify the opponents of Copernicus or Galileo, or deny the tremendous progress which physics owes to the latter. Only he shows that to the extent that both parties in that famous controversy assumed a unique and absolute space they were equally wrong. In this respect, however, Einstein unconsciously brings fresh support to the views of the great Catholic physicist and historian of science, Pierre Duhem.

The greatest triumph of a physical theory is to predict hitherto unsuspected phenomena and to have these predictions experimentally confirmed. This triumph Einstein's theory of gravitation experienced when astronomers during a recent eclipse found that light rays passing near the surface of the sun are deflected just as Einstein predicted. This confirmation, however, by no means proves the whole theory of gravitation—much less his general theory of relativity. The general view that energy has gravity Einstein shares with Max Abraham and others who reject both theories of relativity; and the successful computations as to the course of light rays rest largely on certain independent subsidiary hypotheses. It is highly probable that some future scientist will improve on Einstein's complicated procedure in the theory of gravitation, precisely as Einstein's original paper on relativity improved on the methods of Lorentz and Larmor in the theory of electricity.

It would be absurd to attempt to indicate in the tail end of an article the many philosophical bearings of the theory of relativity. Possibly, however, I may stimulate the reader's own reflection by peremptorily firing at him the following suggestions:

1. The theory of relativity has dealt a death blow—at least so far as scientific physics is concerned—to the view that space and time are empty forms or vessels existing independently of, and possibly prior to, their material contents. Time and space are for physics the correlated numbers or dimensions of material things and events; and in a quite unexpected way our time and space measures have now been shown to be dependent on each other as well as on the material system of which they are aspects. But though every physical system can thus be said to have its own time and

space, the theory of relativity by establishing formulae for correlating all possible physical systems, establishes a universal time and space in the new sense.

2. By showing physical time to be but one aspect of natural events, the theory of relativity reinforces the legitimacy of the great philosophic tradition of viewing things from their eternal aspect. Indeed, Minkowski, one of the most brilliant mathematical minds of modern times, has actually shown how on the basis of the relativity theory the whole of our three-dimensional physics can be viewed as a chapter in a four-dimensional geometry—time being the fourth dimension. To the popular mind the notion of a four-dimensional world has, because of the spiritistic use of it by Zöllner, been associated with irresponsible and unintelligible vagaries. But Minkowski's four-dimensional geometry is a sober, useful and vivid picture of our changing world. Portions of H. G. Wells' *Time-Machine* can give it popular representation. Indeed, whenever we think of any physical event, have we not before us something spread over a time interval as over space?

3. There is a popular philosophic tradition according to which all things are so interconnected that everything makes a difference to everything else. This view is generally fortified by a quotation from Tennyson to the effect that a complete botany of the flower in the crannied wall must include a complete anthropology and theology. Against this view, the original theory of relativity shows that certain motions, while they affect our units of measurement, nevertheless do not affect the final results or laws of nature which we thus obtain, just as figuring the value of your dollars in terms of francs does not change the actual amount in your pocket. You may glorify the unity of the world or the interconnectedness of things as much as you please, but you cannot, without denying the validity of physical science, deny that certain things or aspects of the world are independent of others.

4. It is difficult to determine the precise physical significance which Einstein attaches to his later and more general theory of relativity. You bang your fist on the table, and Einstein shows you how to find a mathematical system of co-ordinates, or time and space elements, in which your fist is defined to have been at rest and the co-ordinates or distances of other objects to have been changing accordingly. This is undoubtedly a great mathematical achievement to the lasting credit of Einstein and his co-worker Grossman, but what bearing, you ask, has it on the physical nature of the world in which we live? Einstein's answer seems to be that the fact of your bringing down your fist is indifferent to the various mathematical descriptions of it, just as it is indifferent as to whether you express it in English or Gaelic. On the other hand, Einstein believes that there are laws of nature and that these laws are expressed by mathematical equations whose essential form is unchanged by a change of co-ordinates or space-time elements which enter into them. If he is justified in asking "What has nature to do with our co-ordinate systems?" why not ask, "What has nature to do with the invariance of our equations?" Might it not be possible to give a physical meaning to the changes of coordinates as well as to the invariance of the equations? In any case it seems a fact that certain mathematical formulae or descriptions serve more effectively than others as keys to all sorts of natural phenomena. If people had kept on saying that the earth is still and the sun in motion would they have made the discoveries which followed the other way of putting it?

Whatever may be the fate of the theory of relativity it has undoubtedly opened up new regions of thought by suggesting new possible connections between fundamental ideas like energy, space, matter, and gravity; and can there be any greater service to the human mind than this opening up of new fields?

from

The Philosophic Meaning of the Copernican Revolution*

by

PHILIPP FRANK

[Philipp Frank, one of the leading contemporary philosophers of science, has been among the first of many who have remarked an interesting parallel between the opposition offered by the established science of the sixteenth century to the innovations of Galileo and Newton, and the opposition offered in recent years by the Galileo-Newtonian traditionists to the innovations of Max Planck and Einstein. Asserting that the Galileo "case" needs to be thoroughly re-examined in the light of recent developments in natural science, Frank investigates the process whereby new open-minded science tends gradually to harden with age until it becomes the very sort of dogmatic closed-minded science it originally opposed. As a student of the work of Pierre Duhem [see the following selection] Frank is aware of the fact that Galileo's science contained from its very inception a dogmatic confidence in the truth of its findings which contrasts sharply with the doctrinal modesty of the new physics, but Frank's argument is, nevertheless, quite impressive.]

. . . Only recently, under the violent impact of twentieth-century physics, particularly the theory of relativity, have

---

* Reprinted by permission of the publishers from Philipp Frank, *Modern Science and Its Philosophy*, (Cambridge, Mass.: Harvard University Press, Copyright © 1941-1949), by the President and Fellows of Harvard College.

the eyes of science students been opened and has the meaning of the Copernican revolution become clear.

If we look into a typical textbook or listen to an average teacher, we learn that before Copernicus, men believed in the testimony of their senses, which told them that our earth is at rest, that the planet Jupiter traverses in twelve years a closed orbit on the celestial sphere and that this curve contains twelve loops. Finally, Copernicus recognized the fallacy of this testimony and proved that "in reality" our earth is in motion and that the planet Jupiter traverses "in reality" a smooth circle around the sun as center. Copernicus exposed the illusions of our senses. Human reason thus scored a clear victory in its struggle against naïve sense experience.

This description of Copernicus' achievement seems to me, conservatively speaking, inadequate. The loops traced by the planets are by no means a sort of optical illusion and neither is the immobility of the earth. As a matter of fact, the planet Jupiter actually traverses every year a loop with respect to a system of reference that participates in the actual revolution of the earth. But the same Jupiter traverses just as truly every twelve years a smooth circle with respect to the system of fixed stars. Neither the loops nor the smooth circle are results of our naïve sense experience. They are two different diagrams representing one and the same set of sense observations. Therefore, the interpretation of Copernicus' achievement as a victory of abstract reason over naïve sense experience is hardly justifiable.

However, we meet occasionally a second interpretation which says almost the opposite of the first one. The hard facts of our sense experience became more and more incompatible with medieval philosophy, which had its roots in speculative reasoning rather than in sense observation. Copernicus finally decided to overthrow the obsolete doctrines of Aristotle and Ptolemy and scored a victory for experience in its struggle against pure speculation. As a matter of fact,

Copernicus was not particularly "tough-minded," if we may use the famous phrase of William James to describe the empirical scientist as distinct from the "tender-minded" believer in pure reasoning. No new facts had been discovered by Copernicus, which had forced him to abandon the geocentric doctrine. The astronomical tables calculated from the Copernican system were in no better agreement with the observed positions of the stars than the previous tables.

Therefore we have to start from the fact that the Copernican revolution meant neither a victory of reason over the illusion of our senses nor a victory of hard facts over pure speculation. To be sure, Copernicus invented a new pattern of description for our observations. His genius manifests itself in the beautiful simplicity of this pattern: he replaced loops by geocentric circles.

Copernicus died in 1543. The Roman Holy Office did not utter an official judgment on the Copernican system until 1616, seventy-three years after Copernicus' death. This Roman verdict will give us the best hint about the philosophic meaning of the Copernican revolution. For the verdict considered specifically the philosophic merit of the new system. The Copernican theory was called "philosophically foolish and absurd."

But not even Copernicus' greatest opponents ever doubted that his system meant a great advance in astronomy. The general opinion in these quarters was that the heliocentric system is "astronomically true," or as it was sometimes phrased, "mathematically true," but in any case "philosophically false" or even "absurd."

We have to do here with a conflict between two conceptions of truth. This conflict has existed through the ages and has created quite often a great confusion of mind. This double meaning of truth has never been dramatized so clearly as by the Copernican revolution and its repercussions. To understand and to evaluate this conflict is the great lesson we can learn from the history of the Copernican ideas.

The Judgment of Posterity

The medieval philosopher St. Thomas Aquinas described very distinctly two different criteria of truth:

> There are two ways to prove the truth of an assertion. The first way consists in proving the truth of a principle from which this assertion follows logically. In this way, one proves in physics the uniformity of the motion of the celestial spheres. The second way consists not in proving a principle from which our assertion can be derived but in assuming our assertion tentatively and in deriving results from it which can be compared with our observations. In this way one derives, in astrology, the consequences of the hypothesis of eccentrics or epicycles concerning the motion of celestial bodies. However, we cannot conclude in this way that the same assumptions cannot be derived, perhaps, from a different hypothesis also.

If a statement of astronomy met only the second criterion, the agreement with observed facts, it was termed "mathematically true." Only if it met also the first criterion, that is, if it could be derived from an evident principle, was it recognized to be "philosophically true." Since Aristotle's physics was supposed to be derived from evident principles, to be philosophically true meant practically to be in agreement with Aristotelian physics.

As Copernicus had been anxious lest his system might not be philosophically true in this sense, he feared some hostility on the part of the theologians who were strict believers in Aristotelian philosophy. He looked for advice on how to behave in this situation, and strange as it may seem to us, the Catholic churchman Copernicus asked a Lutheran theologian from Nuremberg how to avoid trouble. The Nuremberg scholar, Osiander, answered him in a letter of 1541:

> As for my part, I have always felt about hypotheses that they are not articles of faith but bases of calculation. Even if they are false, it does not matter much provided that they describe the observed phenomena correctly. . . . It would, therefore, be an excellent thing for you to play up a little this point in your preface. For you would appease in this way Aristotelians and theologians, the opposition of whom you fear.

This advice meant precisely that Copernicus should not claim "philosophic truth" for his system but should be satisfied with a claim for "mathematical truth."

But Copernicus did not like this compromise. He claimed his system to be as philosophically true as the Ptolemaic system, and perhaps even more so. In this way a conflict flared up, the issue of which was a very subtle distinction. Was the Copernican doctrine a true description of the universe or was it merely an hypothesis which served for calculating the positions of the stars? And how did Copernicus himself look upon the question? Most of the scientists of today are accustomed to regard every theory as a working hypothesis only, and would hardly be prepared to give serious thought to that subtle distinction which is rather an issue of the philosophy of science. But if we go a little deeper into the logical structure of science, we have to recognize that, as a matter of fact, every scientific theory, of whatever period, had to meet the two requirements of a "true theory" which were already familiar to Thomas Aquinas. In reality, no theory was accepted merely because it was a good working hypothesis. In every period of the history of science a theory had to be in agreement with the general principles of physics. The physicists of the nineteenth century would hardly have admitted a theory that was in disagreement with the principle of conservation of energy.

For that reason, practically every theory has been a compromise between these two requirements. This is particularly true of the Ptolemaic system. We read and hear frequently that the Ptolemaic system was in agreement with the Aristotelian philosophy and physics. But Copernicus, we are told, disturbed this harmony and advanced a theory that would contradict explicitly the laws of medieval physics. This was certainly not the opinion of the medieval philosophers themselves. One of the basic principles of medieval physics was the law that terrestrial bodies move in rectilinear paths toward or away from the earth while celestial bodies move in circular orbits with the earth as center, but the Ptolemaic system assumes that sun and planets traverse eccentric circles or epicycles the center of which is not the earth. Therefore, the Ptolemaic system could not be regarded as philosophically true, but at most as a hypothesis that might serve as a basis of calculation.

Thomas Aquinas judged the Ptolemaic system as follows:

> The assumptions made by the astronomers are not necessarily true. Although these hypotheses seem to be in agreement with the observed phenomena we must not claim that they are true. Perhaps one could explain the observed motion of the celestial bodies in a different way which has not been discovered up to this time.

The twelfth-century Arabian philosopher Averroes and his school emphasized very strictly the philosophical criterion of truth and declined to ascribe any truth value to the Ptolemaic system. Says Averroes:

> The astronomers start from the assumption that these [eccentric or epicyclic] orbits exist. From this assumption they derive results that are in agreement with our sense observations. But they have not proved by any means that the presuppositions from which they started are, in turn, necessary causes of these observations. In this case, only the observed results are known but the principles themselves are unknown, for the principles cannot be logically derived from the results. Therefore new research work is necessary in order to find the "true" astronomy, which can be derived from the true principles of physics. As a matter of fact, today there is no astronomy at all, and what we call astronomy is in agreement with our computations but not with the physical reality.

The common opinion among philosophers was rather that the true picture of the universe cannot be discovered by the astronomer, who is restricted to finding out what hypotheses are in agreement with observed facts. If different hypotheses meet this requirement, science cannot decide which is true and, as the Jewish medieval philosopher Moses Maimonides puts it: "Man knows only these poor mathematical theories about the heavens, and only God knows the real motions of the heavens and their causes."

It is certain, therefore, that before the Copernican revolution no theory of the motions of the celestial bodies existed that would meet both criteria of truth. There was in every theory a discrepancy between mathematical and philosophic truth. Against this background we have to interpret the famous dedication letter which Copernicus published as a preface to his great book and in which he recommended his work to the good will of Pope Paul III.

Copernicus affirms that he did not advance his new theory of the motions of the heavens in a spirit of opposition against the established doctrine. His only motive was his conviction that there was no established doctrine. The hypothesis of a circular motion of planets around the earth as center did not account for the observed facts, and the hypotheses of eccentrics or epicycles were not in ageement with the general principles of physics which required uniform circular motions around the earth as center. Since no doctrine existed which could be regarded as "true" from the philosophic as well as from the mathematical angle, Copernicus felt free to suggest a new hypothesis assuming the mobility of the earth.

This hypothesis accounted for the observed motions nearly as well as the Ptolemaic theory of epicycles, but removed some of the epicycles. The motions of the planets became now circular orbits around the sun as center, except for the epicycles which were necessary to account for the inequalities in the motion of planets. In any case there were fewer epicycles and more homocentric orbits in the Copernican, than in the Ptolemaic system. Therefore Copernicus claimed that his theory was in some sense nearer to the requirements of Aristotelian physics than was the geocentric system. The Ptolemaic system was a compromise and, as he believed, a better one.

In any case, Copernicus claimed to give in his theory a true picture of the universe—true in every sense of the word.

By a strange coincidence, Copernicus' book was edited by the same Osiander of Nuremberg whose advice Copernicus did not like to follow. We understand now the famous words of the editor's preface, which had been originally ascribed to the author himself but which reflect only the editor's opinion: "The hypotheses of this book are not necessarily true or even probable. Only one thing matters. They must lead by computation to results that are in agreement with the observed phenomena."

While Copernicus tried to achieve the compromise by arguing that his theory is to a large extent in agreement with the principles of Aristotelian physics, Galileo Galilei, in his famous *Dialogue on the Copernican and Ptolemaic Systems of the World* [translated in our text as the *Dialogue on the Great World Systems*], went a good deal further in the overthrow of medieval science. He no longer attempted to reach the compromise by adjusting his working hypotheses to the requirements of the established principles of physics. On the contrary, he ventured to adjust the principles of physics to the best suitable working hypotheses. This meant dropping the bulk of Aristotelian physics and starting a movement in science that led in time to the philosophy of science which we would call today positivism or pragmatism. The two criteria of truth, which were for medieval thinkers like St. Thomas Aquinas two distinct requirements, have fused more and more into one single requirement: to derive the best description of the observed phenomena from the simplest possible principles, while these principles are justified solely by the fact that they permit this derivation.

Galileo's ideas were not brought into a coherent system of propositions until Isaac Newton advanced his celebrated laws of motion in his *Mathematical Principles of Natural Philosophy*. This book appeared in 1687, approximately 150 years after the Copernican revolution. From the Newtonian principles the Copernican doctrines could be logically derived. Therefore, to the believer in these principles, the Copernican system was now true in the full sense of the word, philosophically and mathematically true.

Let us now ask, "What did the Copernican hypothesis look like when it was derived from the Newtonian principles?" It said that the earth is rotating with respect to absolute space and that the planet Jupiter traverses smooth circular orbits with respect to absolute space. But Newton himself was very well aware that "motion relative to absolute space" has, to use P. W. Bridgman's term, no *operational* meaning, that is, that by no physical experiment can the speed of a body in rectilinear motion with respect to absolute space be measured.

Therefore, the Newtonian system of principles is not a logically coherent system within the domain of physics. Newton himself restored logical coherence by enlarging his system of physical statements by the addition of some theologic propositions. As we read in Burtt's book on the *Metaphysical Foundations of Modern Science:*

> Certainly, at least God must know whether any given motion is absolute or relative. The divine consciousness furnishes the ultimate center of reference for absolute motion. Moreover, the animism in Newton's conception of force plays a part in the premises of the position. God is the ultimate originator of motion. Thus real or absolute motion in the last analysis is the expenditure of divine energy. Whenever the divine intelligence is cognizant of such an expenditure the motion so added to the system of the world must be absolute.

Under the influence of the spirit of the eighteenth century the mixing of theology into science began to be regarded as illegitimate. Strange as it may seem, by the abandonment of theologic argument the Newtonian physics lost logical coherence. Burtt says very correctly: "When in the twentieth century Newton's conception of the world was gradually shorn of its religious relations, the ultimate justification for absolute space and time as he had portrayed them disappeared and the entities were left empty."

Therefore the new principles of physics from which the Copernican theory could be derived were far from being satisfactory. The "philosophic truth" of the Copernican system was still a doubtful thing.

Toward the end of the nineteenth century, Ernst Mach exposed very specifically the logical incoherency of the Newtonian mechanics as a purely physical system. He claimed on good grounds that the principles from which the Copernican system was derived are essentially theologic or metaphysical principles. Mach claimed in the nineteenth century, as Averroes had done in the period of the Ptolemaic systems, that we have no true astronomy, if "true" means "derived from a coherent system of principles of physics."

Mach asked for the removal of the concept of absolute space from physics and for a new physics which contains only terms which have within physics, to speak again with Bridgman, operational meaning.

This program, however, was not carried out until Einstein created his general theory of relativity and gravitation between 1911 and 1915. This theory, as a matter of fact, was the first system of purely physical principles from which the Copernican system of planetary motions could be derived. But the description of these motions looked now very different from the way it had looked as derived from Newton's principles. The concept of absolute space was no longer present. Therefore the statement of the rotation of the earth and of the smooth circular orbits of the planets had now to be formulated quite differently.

From Einstein's principles one could derive the description of the motions of celestial bodies relative to any system of reference. One could demonstrate that the description of the motion of planets becomes particularly simple if one uses the system of fixed stars as a system of reference, but there was still no objection to using the earth as system of reference. In this case, one obtains a description in which the earth is at rest and the fixed stars are in a rotational motion. What appears to be in the Copernican heliocentric system the centrifugal force of the rotating earth becomes in the geocentric system a gravitational effect of the rotating fixed stars upon the earth.

The Copernican system became for the first time in its history not only mathematically but also philosophically true. But at the same moment the geocentric system became philosophically true, also. The system of reference had lost all philosophic meaning. For each astronomical problem, one had to pick the system of reference that rendered the simplest description of the motions of the celestial bodies involved.

The reception of the Einsteinian revolution by the scientists of the twentieth century reminds us in many respects of the reception of the Copernican revolution by the scientists of the sixteenth century. This comparison might help us to understand the philosophical meaning of both.

We may take as an example the way in which Einstein deals with the contraction of moving bodies in the direction of their motion. The verdict of quite a few twentieth-century physicists was: the theory of relativity permits us to derive the observed phenomena from hypothetical principles but it does not give a physical explanation of the contraction. This was an exact repetition of the Roman verdict against the Copernican system. For the meaning was: the theory of relativity may be "mathematically true" but it is certainly "philosophically false." Now "philosophically false" meant not to be in agreement with Newton's principles of physics, while in the sixteenth century the same expression meant not to be in agreement with Aristotle's physics and scholastic philosophy.

But what are the facts affirmed by the Copernican doctrine which are still accepted today as true? Copernicus enthusiastically proclaimed the sun as the center of the universe and said: "In the center of the Universe the sun has its residence. Who could locate this lamp in this beautiful temple in a different or better place than in the center where-from it can illuminate the whole of it simultaneously?"

Even if we restrict the meaning of the "universe" to our galactic system, the Milky Way, this universe is not spherical and the sun is not located in the center. It has been known for a long time that our galactic system has the shape of a lens. Before the distance of very remote stars could be estimated, it was plausible to believe that our sun, with our earth as attendant, is located in the center of the lens. How-ever, in the twentieth century new methods were developed for estimating the great distances of remote stars, in large part by Harlow Shapley and his collaborators of the Harvard Observatory. In particular, Shapley found that our sun is not located near the center of that lens, but approximately 30,000 light years away from it. This means that the sun with our planetary system is near the edge of the lens. According to Copernicus, we inhabitants of the earth have no longer the great satisfaction of being the center of the universe, but we have at least the small satisfaction of being the attendant of a master who has his residence at this center. But accord-ing to Shapley, man has lost all reasons for complacency. He is not even the attendant of a master who occupies the central stage of the universe.

Copernicus probably believed that the orbits of celestial bodies can be described in the best and simplest way by taking the sun as a body of reference. In our twentieth cen-tury, we know that this cannot be true universally. According to Einstein's theory of gravitation, there is no "all-purpose system of reference." Copernicus' suggestion of using the sun is practical only if we restrict ourselves to the motions in our planetary system. For every particular purpose a par-ticular system may be the most suitable.

Copernicus did not discover any new fact that could be regarded as established for all eternity. But he denied to the earth its former role as the only legitimate body of ref-erence, he demonstrated that the sun is the most suitable system for a particular purpose, and he cleared the way for the great new truth that we have complete freedom in our choice of a system of reference.

The Copernican revolution did not end by replacing the earth as master of the universe by the sun or by absolute space, but it was only the first step in a series of revolutions that culminated, so far as we know today, in depicting a democratic order of the universe in which all celestial bodies play an equal part.

from

To Save the Phenomena*

by

PIERRE DUHEM

[Pierre Duhem, author of this final reading, was primarily a theoretical physicist. Yet, as Louis De Broglie, Nobel Prize-winning physicist, has written: "Apart from his strictly scientific works which were brilliant indeed, notably in the domain of thermodynamics, he acquired an extremely extensive knowledge of the history of the physico-mathematical sciences and, after having given much thought to the meaning and scope of physical theories, he shaped a very arresting opinion concerning them, expounding it in various forms in numerous writings." The philosopher Ernst Cassirer has credited Duhem with having carried to rigorous fulfillment the philosophic efforts of Mach, Hertz, Poincaré, and others, to establish an autonomous basis for physical theory, purging it of the last residues of burdensome metaphysics. Philipp Frank, author of the preceding reading, has hailed him as "the greatest and most accurate student of the history of physics."

The chief monument of historical research left to us by Pierre Duhem is his massive *Le Système du Monde, Histoire des Doctrines Cosmologiques de Platon à Copernic,* in ten volumes, five published posthumously, the last in 1959. This thoroughly documented work reviews the history of astronomy and cosmology from the ancient Greek, through the Roman,

---

* Translated by Anne Paolucci, from Pierre Duhem, *Sozein ta Phainomena, Essai sur la Notion de Théorie Physique de Platon à Galilée.* Copyright © 1961 by Anne Paolucci.

early Christian, Arabic, and medieval periods, to the time of Copernicus, tracing the development through all its evolutions, revolutions, and involutions, showing how the science of each epoch is nourished by the systems of past centuries, indicating how often the scientific certainties of one proud age are laughed at as absurdities in the next, while as often today's absurdities are seen forcing themselves forward into acceptance as the high truths of tomorrow. A master of Latin and Greek philology and a competent paleographer, Duhem resorted in his researches as often as possible to original materials, and was able thereby to correct many of the traditional errors perpetuated from generation to generation by historians of science who work only from published documents and other secondary sources.

The mass of erudition suporting *Le Système du Monde* is so enormous as to appear overwhelming on first acquaintance. Duhem, however, drew out its central thread of meaning in advance, for separate publication in a series of short essays under the Greek title *Sozein ta Phainomena* (To Save the Phenomena): *Essai sur la Notion de Théorie Physique de Platon à Galilée,* the final portion of which is presented here.

One of the main purposes of all Duhem's historical writings and the chief object of his *Sozein ta Phainomena* was to provide a safeguard in the future against the two most serious derangements—"the mad ambition of dogmatism as well as the despair of Pyrrhonian skepticism"—to which the minds of physicists, in the past, have been especially prone. His work is, therefore, particularly important today when many of the leading contemporary physicists are pressing the essential skepticism of the relativist and indeterminist positions to its extremest consequences, driving the last defenders of Galilean and Newtonian dogmatism from the field. The special relevance of Duhem's work in this area was indicated, the reader will recall, by Professor Morris R. Cohen, in his discussion of the philosophical implications of Einstein's later theory of relativity, which, as he says, "brings fresh support to the views of the great . . . physicist and historian of science, Pierre Duhem."]

Astronomical hypotheses are mere artifices designed to "save" phenomena; provided they fulfill this end, they need not be true or even probable.

This view seems to have been generally accepted, by astronomers and theologians from the time of the publication of Copernicus' book and Osiander's preface, to that of the Gregorian reform of the calendar. On the contrary, in the course of the half-century that extends from the reform of the calendar to the condemnation of Galileo, we find it relegated to obscurity, indeed, even violently opposed in the name of a widespread realism that seeks to find in astronomical hypotheses affirmations regarding the nature of things, and which requires henceforth that these hypotheses accord with the doctrines of Physics and with the texts of Scripture. . . .

J. Kepler is, beyond any doubt, the most convinced and the most illustrious representative of this new realism.

In the preface itself of his first work, the *Mysterium cosmographicum*, printed in 1596, Kepler informs us that six years earlier, at Tubingen, as assistant to Michel Maestlin, he had already been captivated by the system of Copernicus: "From that moment, I was intent upon attributing to the Earth not only the movement of the outermost movable sphere but also the solar movement; and while Copernicus attributed these motions to the earth for mathematical reasons, I would attribute them for physical, or, if you prefer, metaphysical reasons."

Kepler is Protestant, but profoundly religious; he would not regard the Copernican hypotheses as conformable to reality if they were contradicted by Holy Scripture; before entering upon the terrain of Metaphysics or Physics, therefore, he must traverse that of Theology. "From the outset of this discussion on nature," so he writes at the beginning of Chapter I of his *Mysterium cosmographicum*, "we must take care to say nothing contrary to Holy Scripture."

Thus Kepler indicates for the Copernicans the way they will thereafter be obliged to follow: as realists, they require that their hypotheses conform to the nature of things; as Christians, they acknowledge the authority of the sacred Text; hence they are led to reconcile their astronomical doctrines with Scripture, and are constrained to assume the role of theologians.

They might have avoided such constraint had they thought of astronomical hypotheses what Osiander thought of them; but those who pursued faithfully the indications of Copernicus and Rhaeticus could not endure the doctrine expressed in the famous preface. "Some persons," said Kepler, "make much of the example of an exceptional demonstration in which, from false premises, by means of a rigorous syllogistic deduction, one is able to draw a true conclusion; on the strength of this example they try to prove that the hypotheses entertained by Copernicus might be false and that, nevertheless, true phenomena could derive from them as from their proper principles; I have never been able to accept such a view. . . . All that Copernicus discovered *a posteriori*, all that he demonstrated by observation, can, I do not hesitate to assert, be demonstrated *a priori*, by means of geometric axioms, in a manner that would overcome all hesitation and that would even win the approval of Aristotle himself, were he still alive. . . ."

Kepler is not content to criticize the doctrine upheld by Osiander. . . ; he means, further, to put into practice that realism the principles of which he had set down; evidence of this realism is supplied us by that greatest of the works which his genius produced, the *Epitome Astronomiae Copernicanae*.

Its realism is manifest from the opening of the first book of the work: "Astronomy," says Kepler, "is a part of Physics," and the importance of this aphorism is at once revealed in what the author tells us *De causis hypothesium:* "The third part of the astronomer's 'baggage' is Physics; generally it is not considered necessary for the astronomer; and yet the Science of the astronomer has a great bearing on the object to be attained by this part of Philosophy which, without the astronomer, would remain incomplete. Indeed, astronomers should not be allowed absolute license to assume anything at all, without sufficient reason. You should be able to offer likely reasons for the hypotheses which you assume to be the true causes of phenomena; you should, therefore, search for the foundations of your astronomy in a higher science, that is to say, in Physics or in Metaphysics; furthermore, with the geometric, physical or metaphysical arguments supplied by your particular science to support you, you are not prevented, in turn, from moving beyond the limits of this science to discover objects pertaining to these higher doctrines."

In the course of his *Epitome,* Kepler takes every possible occasion to support his hypotheses with arguments supplied to him by Physics and Metaphysics. What a Physics and what a Metaphysics! But this is hardly the place to describe what strange dreams, what childish fancies, Kepler designated by these two names. We do not have to examine how Kepler constructed his astronomy; we have to know, simply, how he wanted it to be constructed. And, we now know, he wanted the science of celestial motions to rest on foundations guaranteed by Physics and by Metaphysics; he insisted that astronomical hypotheses be in no way contradicted by Scripture.

We find, moreover, a new ambition manifesting itself in Kepler's writings: Founded on true hypotheses, Astronomy can, by means of its conclusions, contribute to the progress of the Physics and Metaphysics which have supplied it with its principles. . . .

When Galileo accepted the Copernican system, he did so in the same spirit. . . ; he wanted the hypotheses of the new system to be, not artifices designed for the calculation of tables, but propositions conforming to the nature of things; he wanted them to be established on the grounds of Physics. One might, indeed, say that physical confirmation of the Copernican hypotheses is the center toward which all, even the most diverse, of Galileo's researches tend; his observations as an astronomer, his theories as a pioneer in mechanics converge toward this same end. Further, because he insisted that the foundations of Copernican astronomy be truths, and because he did not believe that a truth could contradict Scripture, which he acknowledged to be of divine inspiration, he was bound to attempt to reconcile his assertions with biblical texts; when the time arrived for him, he turned theologian; his celebrated letter to Marie-Christine of Lorraine bears us out.

In claiming that the hypotheses expressed physical truths, in declaring that they did not seem to him to contradict the Holy Scriptures, Galileo was, like Kepler, wholly in the tradition of Copernicus and Rhaeticus. He set himself against those who represented the tradition of Tycho Brahe, the Protestant, and the Jesuit Rodolphe Clavius. What these had said around the year 1580 the theologians of the Holy Office solemnly proclaimed in 1616.

They seized on these two fundamental hypotheses of the Copernican system:

Sol est centrum mundi et omnino immobilis motu locali;
Terra non est centrum mundi nec immobilis, sed secundum se totam movetur, etiam motu diurno.

They asked themselves whether or not these two propositions had the two characteristics which, by common accord, Copernicans and Ptolemains alike required of all admissible astronomical hypotheses: Were these propositions compatible with sound Physics? Were they reconcilable with divinely inspired Scripture?

Now, for the Inquisitors, sound Physics was the Physics of Aristotle and Averroes; it laid down for them, unequivocally, the answer that they were to give to the first question: the two hypotheses challenged were *stultae et absurdae in Philosophia.*

As for Scripture, the consultants of the Holy Office refused to accept any interpretation of them which was not supported by the authority of the Fathers; the answer to the second question consequently forced itself upon them: The first proposition was *formaliter haeretica,* the second was *ad minus in fide erronea.*

The two propositions censored had neither one nor the other of the two characteristics which were supposed to distinguish all admissible astronomical hypotheses; they were to be wholly rejected, therefore, not to be used even for the sole purpose of *saving the phenomena;* thus the Holy Office prohibited Galileo from teaching the Copernican doctrine *in any fashion.*

The condemnation carried through by the Holy Office was the result of a clash between two "realist" positions; this violent opposition might have been avoided, the debate between the Ptolemaics and the Copernicans might have been limited to the field of Astronomy, if the wise precepts concerning the nature of scientific theories and the hypotheses on which they rest had been heeded; these precepts, formulated by Posidonius, Ptolemy, Proclus, Simplicio, had come down directly, through an uninterrupted tradition, to Osiander, to Reinhold, to Melanchthon; but they now seemed to be wholly forgotten.

And yet there were at hand voices of authority to call attention to them once again.

One of these voices was that of Cardinal Bellarmine, the same who in 1616 was to examine the Copernican writings of Galileo and Foscarini; as early as April 12, 1615 Bellarmine had written Foscarini a letter full of good sense and prudence from which some passages are here given:

"It seems to me that Your Reverence and Galileo will act prudently by contenting yourselves to speak *ex suppositione* and not in absolute fashion, as I have always believed Copernicus to have spoken. It is well to say that in assuming the Earth to move and the sun to remain stationary one can "save" all the appearances better than can be done with eccentrics and epicycles; there is no danger in that and it suffices for the mathematician. But to want to affirm that the sun really remains stationary at the center of the universe, that it turns only upon itself without moving from east to west, that the Earth occupies the third sky and that it turns with great speed around the Sun, is very perilous; it is likely not only to irritate all the philosophers and all the scholastic theologians, but also to harm the faith and to render false Holy Scripture. . . .

"If it could be demonstrated with certainty that the Sun stands at the center of the Universe, that the Earth is in the third sky, that it is not the Sun that turns around the Earth but the Earth that turns around the Sun, then we should have to proceed with great circumspection in the explication of the Scriptures. . . . But I shall not believe that such a demonstration is possible until it can be shown to me. It is one thing to prove that one can 'save' appearances by supposing that the Sun is at the center of the Universe and that the Earth is in the heavens, but quite another thing to demonstrate that the Sun is really at the center of the Universe and the Earth really in the heavens. Regarding the first demonstration, I believe it can be given; but regarding the second, I strongly doubt it; and in a case of simple doubt, you must not abandon Scriptures as the Holy Fathers have expounded them. . . ."

Galileo knew of the letter addressed by Bellarmine to Father Foscarini; several writings edited between the time when he learned of this letter and his first condemnation contain answers to the Cardinal's arguments; examination of these writings, from which excerpts were first published by M. Berti, enables us to grasp the vital thought of Galileo regarding astronomical hypotheses.

One piece written toward the end of the year 1615 and addressed to the consultants of the Holy Office warns them against two errors: the first is to claim that the mobility of the Earth is, in some way, *a great paradox* and *a manifest piece of folly,* which has not been proved as yet and which can never be proved. The second is to believe that Copernicus and the other astronomers who have assumed this mobility "did not believe that it was true in fact and in nature," that they have only admitted it as a supposition in order to account for the appearance of the celestial movements more easily, in order to make astronomical calculations more convenient.

In affirming that Copernicus believed in the reality of the hypotheses formulated in *De Revolutionibus;* in proving by an analysis of this work that Copernicus did not admit the mobility of the Earth and the immobility of the Sun only *ex suppositione,* as Osiander and Bellarmine would have it, Galileo was upholding historical truth. But what interests us more than his judgment as an historian is his opinion as a physicist. Now this is easily inferred in the piece we are analyzing. Galileo thought that the reality of the movement of the Earth is not only demonstrable but already demonstrated.

This thought emerges even more clearly in another text; there, not only do we see that Galileo thought the Copernican hypotheses could be demonstrated, but we learn further how he understood the demonstration to have been carried out:

"To refuse to believe that the movement of the Earth is susceptible of demonstration so long as such demonstration has not actually been provided is to act very prudently; we do not, therefore, expect anyone to believe such a thing without demonstration; we would expect only that for the good of the Holy Church one examine with extreme severity all that those who maintain such a doctrine have produced or might produce; that none of their assumptions be admitted unless the arguments that sustain it greatly exceed the arguments of the other side; that their judgment be rejected unless supported by 90 percent of the arguments. But in turn, when it shall have been proved that the opinion advanced by the philosophers and astronomers of the opposite side is unquestionably false, that it is of absolutely no weight then one ought not to scorn the opinion of the first side, one ought not to consider it so paradoxical as to be forever beyond the possibility of clear demonstration. We may reasonably propose such broad conditions for this dispute; in fact, it is clear that those holding to the side of error cannot support themselves with any argument or any experience that is valid: on the contrary, on the side of truth everything must harmonize and agree.

"It is certainly not the same thing to show that by assuming the Sun as stationary and the Earth as moving, appearances are saved, and to demonstrate that such hypotheses are really true in Nature; but what is otherwise and much more true is that with the commonly accepted system these appearances cannot be accounted for, so that this system is unquestionably false; similarly it is clear that the system which accords very closely with appearances can be true, and, in a given situation, one neither can or ought to seek another or greater truth than that which answers for all the particular appearances."

If one were to press this last proposition just a bit further, one could easily draw from it the doctrine that Osiander upheld, that Bellarmine upholds; that is to say, precisely the doctrine which Galileo attacks. Logic thus constrains the great Pisan geometer to formulate a conclusion directly contrary to that which he thought to establish. But in the lines that precede, his thought is clearly apparent.

The pending dispute appears to his mind's eye as a sort of duel. We are confronted with two doctrines, each of which claims to be in possession of the truth; but one speaks the truth, the other lies; who will decide? Experience. That doctrine of the two with which experience will refuse to accord will be judged erroneous and, by the same token, the other doctrine will be declared to accord with reality. The refutation of one of these two conflicting systems guarantees the certainty of the opposing system, just as, in geometry, the absurdity of one proposition carries with it the certitude of the contradictory proposition.

If anyone doubts that Galileo really held the opinion which we attribute to him on the subject of the proof of an astronomical system, he will be convinced of it, we believe, by reading the following lines:

"The most expedient and surest way to show that the position of Copernicus is not contrary to Scripture would be, as I see it, to show by a thousand proofs that this proposition is true and that the contrary position cannot in any way stand; consequently, since two truths cannot contradict one another, it must follow that the position admitted to be true accords with Holy Scriptures."

On the value of the experimental method and the art of using it, Galileo has pretty much the same opinion as that which Francis Bacon will formulate; he conceives the proof of a hypothesis in imitation of the demonstration by absurdity used in geometry; by condemning one system as erroneous, the experiment confers certitude upon the opposing system; positivistic science advances by a series of dilemmas, each of which is resolved with the help of an *experimentum crucis.*

This manner of conceiving the experimental method was destined to enjoy great vogue, for it is very simple; but it is entirely false, for it is too simple. Admit that the phenomena cease to be saved by the Ptolemaic system; that system would certainly have to be considered false. It will not in the least follow that the Copernican system is true, for the Copernican system is not purely and simply the contrary of the Ptolemaic system. Admit that the Copernican hypotheses succeeded in saving all known appearances; one will conclude from this that these hypotheses may be true; one will not conclude that they are certainly true; to justify that conclusion it would first have to be proved that no other combination of hypotheses were conceivable that could save appearances just as well; and this last demonstration has never been given. In Galileo's own time, was it not possible to save all the observations that could be mustered in favor of the Copernican system just as well by the system of Tycho Brache?

These remarks were made often enough in Galileo's time. Their truth had flashed upon the eyes of the Greeks the day when Hipparchus succeeded in saving the solar movement as well by means of an eccentric as by means of an epicycle; St. Thomas Aquinas had formulated them with the greatest clarity; Nifo, Osiander, Alexander Piccolomini, Giuntini, had repeated them after him. Once again an authoritative voice was to remind the illustrious Pisan of them.

Cardinal Maffeo Barberini, who was soon to be elevated to the papacy under the name Urban VIII, had a meeting with Galileo, after the 1616 condemnation, to discuss the Copernican doctrine; Cardinal Oregio, present at this meeting, has left us an account of it; in this meeting the future pope, by means of arguments similar to those which we have just recalled, laid bare the hidden error of this Galilean argument: since all the celestial phenomena accord with the Copernican hypotheses while they are not saved by the Ptolemaic system, the Copernican hypotheses are certainly true; hence they are of necessity in accord with Holy Scripture.

According to Oregio's account, the future Urban VIII advised Galileo "to note carefully whether or not there is agreement between the Holy Scriptures and what he had conceived regarding the movement of the Earth, for the purpose of saving the phenomena that are manifest in the sky and all that philosophers commonly regard as settled, by means of observation and a minute examination in what has to do with the movements of the sky and the planets. Granting, in effect, everything this great scientist had conceived, he asked him if it were beyond the power and wisdom of God to arrange and move the orbs and planets in another way, and do this, however, in such a way that all the phenomena manifest in the skies, all that is taught concerning the movements of the stars, their order, their position, their distance, their arrangement, still could be saved.

"If you mean to say that God cannot or knows not how to do this, you must prove," added the holy prelate, "that all this could not, without involving contradiction, be obtained by a system other than the one you have conceived; God, indeed, is capable of all that does not lead into contradiction; moreover, since God's science is not inferior to his power, if we say that God could do it, we must also say that he could know it.

"If God knew how and was able to arrange all things in a way other than what you have imagined, and this in such a way that all the results enumerated were still saved, we are not in the least obliged to reduce divine power and wisdom to this system which you have conceived.

"Having heard these words, the great scientist remained silent."

The man who was to become Urban VIII had reminded him clearly of this truth: the confirmations of experience, howsoever numerous and precise they might be, could never transform a hypothesis into a certitude, for it would be necessary, in addition, to demonstrate this proposition: that the same facts of experience would, necessarily, contradict all other hypotheses that might be conceived.

Were these very logical and reasonable admonitions of Bellarmine and Urban VIII sufficient to convince Galileo, to sway him from his exaggerated confidence in the scope of the experimental method, in the worth of astronomical theories? We may well doubt it. In his celebrated *Dialogue* of 1632 on the two great systems of the World, he asserts from time to time that he treats the Copernican doctrine as a pure astronomical hypothesis, without claiming it to be true in nature; these protests, which contradict the proofs accumulated by Salviati, one of the speakers, to sustain the reality of the Copernican positions, are undoubtedly nothing more than pretexts to break the interdiction laid down in 1616. At the very moment when the dialogue is about to end, Simplicio, the peripatetic butt and target, to whom is assigned the thankless task of defending the Ptolemaic system, concludes with these words:

"I confess that your thought seems to me much more ingenious than many of those I have had occasion to hear; even so, I do not hold it true, or conclusive; in fact, I keep always before my mind's eye a very solid doctrine which I received from a very learned and eminent person and before which we must pause. Indeed, I would like to ask you both this question: Can God with his infinite power and his infinite science give to the element of water the oscillating movement that we observe, in any other way than by making the containing vessel move? . . . If the answer is yes, I conclude at once that it would be foolhardy to want to limit and constrain divine wisdom and power to one particular conjecture alone."——"An admirable and truly angelic doctrine," Salviati answers; "One can answer, in a way that agrees just as well, by means of another doctrine which is divine: although he allows us to argue about the constitution of the World, God adds . . . that we are in no condition to discover the work which his hands have fashioned."

Perhaps through the mouth of Simplicio and of Salviati, Galileo wished to address a delicate piece of flattery to the Pope; perhaps he also wished to answer the old argument of Cardinal Maffeo Barberini with a touch of ridicule; Urban VIII took it in this guise: against the impenitent realism of Galileo he gave full license to the intransigent realism of the peripatetics of the Holy Office; the condemnation of 1633 confirmed the sentence of 1616.

Conclusion

Many philosophers since Giordano Bruno have reproached Andreas Osiander harshly for the preface which he added to the book of Copernicus. The recommendations made to Galileo by Bellarmine and Urban VIII have been treated with hardly less severity, since the day when they were first published. Physicists of our own time have weighed more minutely than their predecessors the exact value of the hypotheses employed in astronomy and physics; they have witnessed the dispelling of many illusions, which not long since still passed for certitudes; it is their duty, today, to acknowledge and to declare that logic was on the side of Osiander, Bellarmine and Urban VIII, and not on the side of Kepler and Galileo; that the former grasped the precise import of the experimental method and that the latter were mistaken in its regard.

The history of science, however, honors Kepler and Galileo, whom it ranks among the great reformers of the experimental method, whereas it does not so much as mention the names of Osiander, Bellarmine or Urban VIII. Is this the height of injustice on its part? May it not be true, on the contrary, that those who attributed to the experimental method a false import and an exaggerated value worked more effectively and contributed much more toward the perfecting of this method than those who, from the outset, had a more precise and just appreciation of it?

The Copernicans plunged headlong into an illogical realism, when everything seemed to be pressing them to avoid such error, when by attributing to astronomical hypotheses the correct value determined by so many men of authority it would have been easy for them to avoid both the quarrels of the philosophers and the censures of the theologians. Such strange behavior calls for an explanation! But can it be explained otherwise than by the attraction of some great truth, a truth perceived too vaguely by the Copernicans for them to formulate it in its purity, to extricate it from the erroneous affirmations beneath which it concealed itself; but a truth so keenly felt that neither logical precepts nor practical counsels could weaken its invisible attraction. What was this truth? That is what we must now attempt to define.

Through antiquity and the Middle Ages, Physics appears to us made up of two parts, one so completely distinct from the other that they are, so to speak, opposed to one another; on one side we find the Physics of celestial and imperishable things, on the other the Physics of sublunary things subject to generation and decay.

Those things which the first of the two Physics deals with are reputed to be of an infinitely higher nature than those which the second deals with. From this one concludes that the first is incomparably more difficult than the second; Proclus teaches that sublunary Physics is accessible to man, whereas celestial Physics transcends him and is reserved for divine Intelligence; Maimonides shares this opinion of Proclus; according to him, celestial Physics is full of mysteries, knowledge of which God has reserved for Himself alone, whereas terrestrial Physics is to be found, fully worked out, in the work of Aristotle.

Contrary to what the men of antiquity and of the Middle Ages thought, the celestial Physics which they had constructed was singularly more advanced than their terrestrial Physics.

From the age of Plato and Aristotle, the science of the stars was organized according to the plan which we today still impose on the study of nature. On one hand there was Astronomy; geometers like Eudoxus and Callipus formed mathematical theories by means of which they could describe and predict celestial movements, while observers noted the degree of correspondence between the mathematical predictions and natural phenomena. On the other hand there was

Physics proper, or to use the modern terminology, celestial Cosmology; thinkers like Plato and Aristotle speculated on the nature of the stars and on the cause of their movements. How were these two parts related to one another? What precise line of division was there between them? What affinity united the hypotheses of the one with the conclusions of the other? These are questions which astronomers and physicists discuss throughout antiquity and the Middle Ages, which they resolve in different ways, because their minds are guided by diverse tendencies very similar to those which appeal to modern men of science.

Much was required before the Physics of sublunary things would attain in its own good time the same degree of differentiation and organization. It, too, in modern times will divide itself in two parts very similar to those into which celestial physics from antiquity was divided. In its theoretic part it will combine mathematical systems which will reveal by their formulas the precise laws of the phenomena. In its cosmological part it will seek to grasp the nature of corporeal things, of their attributes, of the forces to which they respond or which they exert, of the combinations they can form among themselves.

During antiquity, during the Middle Ages and the Renaissance, it was difficult to make this division; sublunary Physics was hardly aware of mathematical theory. Two subdivisions of this physics, Optics or *Perspective* and statics or *Scientia de ponderibus,* had alone assumed this form, and physicists were greatly embarrassed when they wanted to assign to *Perspective* and to the *Scientia de ponderibus* their rightful places in the hierarchy of sciences. Except for these two subdivisions, the analysis of the laws which regulate phenomena continued to be purely qualitative, lacking precision; it had not yet freed itself from cosmology.

In Dynamics, for example, the laws of free-falling bodies, dimly perceived since the fourteenth century, the laws of the movement of projectiles, vaguely surmised in the sixteenth century, remained entangled in the metaphysical discussions on local movement, on natural movement and forced movement, on the coexistence of mover and movable. Only in Galileo's day, at the very time that its mathematical character was becoming more precise, do we see the theoretic part free itself from the cosmological part. Until then, the two parts had remained intimately united, or rather, entangled in an inextricable manner. Their aggregate constituted the Physics of local movement.

On the other hand, the old distinction between the Physics of celestial bodies and the Physics of sublunary things underwent a gradual effacement. Following Nicholas of Cusa, following Leonardo da Vinci, Copernicus had dared to regard the Earth as one of the planets. By his study of the star that had appeared then disappeared in 1572, Tycho Brahe had shown that the stars too could come into being and pass away. In discovering the sunspots and the mountains of the moon, Galileo brought to completion the union of the two Physics into a single science.

Consequently, when a Copernicus or a Kepler or a Galileo declared that astronomy should take as its hypotheses propositions the truth of which has been established by Physics, that assertion, seemingly one, included in reality two quite distinct propositions.

Such an assertion, in fact, could mean that the hypotheses of astronomy were judgments on the nature of celestial things, and on their actual movements; it could mean that in regulating the correctness of these hypotheses, the experimental method enriched our cosmological knowledge with new truths. This first meaning is found, so to speak, on the very surface of the assertion; it is immediately apparent; it is this meaning which the great astronomers of the sixteenth and seventeenth centuries saw clearly, it is this same one which compelled their allegiance. Now, taken with this meaning, their assertion was false and harmful; Osiander, Bellarmine and Urban VIII regarded it, justifiably, as opposed to logic; but this assertion was to breed countless errors before the decision to reject it was finally reached.

Underlying this first sense, illogical but obvious and seductive, the assertion of the Renaissance astronomers contained another; in demanding that astronomical hypotheses accord with the teachings of physics, they were in effect demanding that the theory of celestial movements rest on foundations that could also support the theory of the movements we observe here below; they were demanding that the course of the stars, the ebb and flow of the sea, the movement of projectiles and falling bodies be saved with one and the same set of postulates formulated in the language of mathematics. Now this sense remained deeply hidden; neither Copernicus nor Kepler nor Galileo saw it clearly; it remained, however, disguised yet fruitful underneath the obvious but erroneous and dangerous sense which alone these astronomers seized. And while the false illogical meaning which they attributed to their principle brought forth polemics and quarrels, it was the true but hidden meaning of this same principle which gave birth to the scientific endeavors of these pioneers; while they were striving to maintain the correctness of the first sense, they were moving unconsciously toward establishing the accuracy of the second sense: while Kepler was multiplying his attempts to account for the movements of the stars with the aid of the properties of the currents of water or of magnets, while Galileo was endeavoring to accord the paths of projectiles with the movement of the earth or to derive from this last movement an explanation of the tides, both of them thought they were proving that the Copernican hypotheses were founded in the very nature of things; but the truth which they were introducing little by little into the sphere of science is that one and the same Dynamics should, by means of a single set of mathematical formulas, represent the movements of the stars, the oscillations of the ocean and the fall of heavy bodies; they thought they were correcting Aristotle; they were preparing for Newton.

In spite of Kepler and Galileo, we today hold with Osiander and Bellarmine that the hypotheses of physics are but mathematical artifices designed to *save the phenomena;* but thanks to Kepler and Galileo, we now call upon them to *save, as a single whole, all the phenomena of the inanimate universe.*